ROBO WORLD

WOMEN'S ADVENTURES IN SCIENCE

ROBO WORLD
the story of robot designer
CYNTHIA BREAZEAL

by Jordan D. Brown

Franklin Watts
A Division of Scholastic Inc.
New York • Toronto • London • Auckland • Sydney
Mexico City • New Delhi • Hong Kong
Danbury, Connecticut

Joseph Henry Press
Washington, D.C.

Author's Acknowledgments

When I first contacted Cynthia Breazeal, I panicked to learn that her first child was almost due. How would she find the time to help with this book? Fortunately, she spent many hours being interviewed, suggesting resources, and reviewing the manuscript. Others who know Cynthia also offered enormous help. Norman, Juliette, and William Breazeal, Ben Green, and Bobby Blumofe provided great stories and photos. My thanks go to Rodney Brooks, Rosalind Picard, Sherry Turkle, Colin Angle, Matt Williamson, and Anne Foerst for their insights about Cynthia's grad school years. Dan Stiehl gave me a terrific tour of the Robotic Life lab. Deborah Douglas and Lijin Aryananda told me about Kismet's retirement to the MIT Museum. Additional thanks go to Brad Ball, Katie Ellinger, Roger Freedman, Kathleen Kennedy, Richard Landon, Lindsay MacGowen, Senia Maymin, Randy Pausch, Scott Senften, John Underkoffler, Claire Velez, and Stan Winston. I thank Jonathan Rosenbloom for recommending me to Joseph Henry Press, as well as Kate Nyquist Jerome, whose sharp editorial eye helped focus the manuscript. Above all, I thank my parents, Eileen and Stephen Brown, who nurtured my love of writing since I was a boy, and my wonderful wife Ellen Reifenberger, who encourages my creative ventures with her love, generosity, and intellectual curiosity. —JDB

Cover photo: Roboticist Cynthia Breazeal stands in front of an enlarged image of Kismet, the autonomous robot she designed.

Cover design: Michele de la Menardiere

Library of Congress Cataloging-in-Publication Data

Brown, Jordan.
   Robo world : the story of robot designer Cynthia Breazeal / Jordan Brown.
        p. cm. — (Women's adventures in science)
   Includes bibliographical references and index.
   ISBN 0-531-16782-8 (lib. bdg.) ISBN 0-309-09556-5 (trade pbk.) ISBN 0-531-16957-X (classroom pbk.)
   1. Breazeal, Cynthia L. 2.   Robotics. 3.   Computer engineers—United States—Biography. I. Title. II. Series.
   TJ211.B699 2005
   629.8'92'092—dc22

                                                                    2005000826

Any opinions, findings, conclusions, or recommendations expressed in this volume are those of the author and do not necessarily reflect the views of the National Academy of Sciences or its affiliated institutions.

Printed in the United States of America.
1 2 3 4 5 6 7 8 9 10 R 14 13 12 11 10 09 08 07 06 05

ABOUT THE SERIES

The stories in the *Women's Adventures in Science* series are about real women and the scientific careers they pursue so passionately. Some of these women knew at a very young age that they wanted to become scientists. Others realized it much later. Some of the scientists described in this series had to overcome major personal or societal obstacles on the way to establishing their careers. Others followed a simpler and more congenial path. Despite their very different backgrounds and life stories, these remarkable women all share one important belief: the work they do is important and it can make the world a better place.

Unlike many other biography series, *Women's Adventures in Science* chronicles the lives of contemporary, working scientists. Each of the women profiled in the series participated in her book's creation by sharing important details about her life, providing personal photographs to help illustrate the story, making family, friends, and colleagues available for interviews, and explaining her scientific specialty in ways that will inform and engage young readers.

This series would not have been possible without the generous assistance of Sara Lee Schupf and the National Academy of Sciences, an individual and an organization united in the belief that the pursuit of science is crucial to our understanding of how the world works and in the recognition that women must play a central role in all areas of science. They hope that *Women's Adventures in Science* will entertain and enlighten readers with stories of intellectually curious girls who became determined and innovative scientists dedicated to the quest for new knowledge. They also hope the stories will inspire young people with talent and energy to consider similar pursuits. The challenges of a scientific career are great but the rewards can be even greater.

Other Books in the Series

Beyond Jupiter: The Story of Planetary Astronomer Heidi Hammel

Bone Detective: The Story of Forensic Anthropologist Diane France

Forecast Earth: The Story of Climate Scientist Inez Fung

Gene Hunter: The Story of Neuropsychologist Nancy Wexler

Gorilla Mountain: The Story of Wildlife Biologist Amy Vedder

Nature's Machines: The Story of Biomechanist Mimi Koehl

People Person: The Story of Sociologist Marta Tienda

Space Rocks: The Story of Planetary Geologist Adriana Ocampo

Strong Force: The Story of Physicist Shirley Ann Jackson

Contents

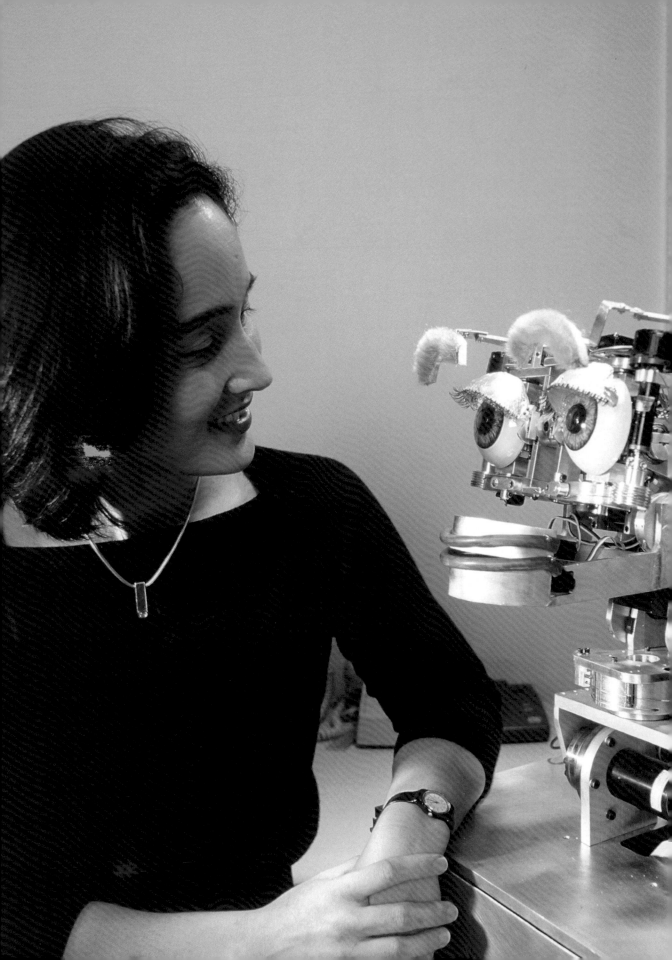

Creature Creator

When you hear the word "robot," you might think of futuristic machines that battle each other in the movies or furry toy pets that answer commands. But there's another exciting world where robots do more than entertain. Cynthia Breazeal lives and thrives in that world. She is a roboticist, a scientist who designs, programs, and experiments with robots.

Cynthia's mission is to build innovative robots that can work and learn in cooperation with people—but not as servants or tools. She wants robots to make our lives better through their abilities, social skills, and even "personalities." By creating life-like machines, Cynthia is also making interesting discoveries about our own human behavior.

Cynthia has applied her engineering and computer programming skills to incredible robotics projects: Attila, Hannibal, Cog, Kismet, and Leonardo are known throughout the world. They are also examples of Cynthia's strong creative talent that allows her to combine art and science in groundbreaking ways. To many, this makes her a visionary—a person who can imagine things that don't yet exist.

How did Cynthia become a world-famous roboticist? What challenges did she face along the way? As you read her story, you'll soon see how Cynthia's curiosity, creativity, and competitive spirit helped her achieve her dreams.

It was *a* time
when her **robot** pal

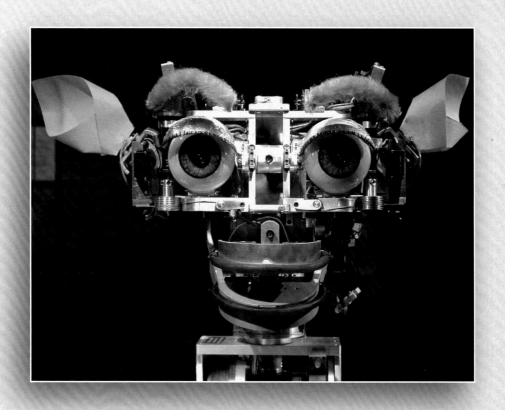

seemed almost
alive.

VISITING AN OLD FRIEND

Since April 2003, a cartoonish-looking robot named Kismet has been one of the star attractions of an exhibit entitled "Robots and Beyond: Exploring Artificial Intelligence." Most people who view this display at the Massachusetts Institute of Technology (MIT) are amazed and awed by the idea of a "lifelike" robot. They are also impressed by the videos of Kismet in action. But one visitor looks at Kismet in a very different way.

Cynthia Breazeal feels an odd mix of pride, melancholy, and nostalgia when she stops by the exhibit. Why? Kismet was one of the robots Cynthia created as a graduate student at MIT. And although she is thrilled and honored that Kismet is such an important part of the exhibit, a wistful feeling hits Cynthia when she realizes that, in many ways, Kismet is no more.

All that remains of Kismet is its head and neck. (Okay, it never actually had a body.) Kismet's brilliant "brain" is gone. That's because the 15 networked computers that once ran the robot's motors, sensors, and programs belong to MIT's Computer Science and Artificial Intelligence Lab (CSAIL). Other graduate students have commandeered these computers to run software for their own robot projects. As a result, Kismet is no longer an active robot.

Cynthia Breazeal shows her silly side at MIT in May 2000 *(above)*. Hours before, she had defended her thesis research about a robot named Kismet *(opposite)*.

1

The MIT Museum visitors who stare at Kismet's motionless face might wonder why the *Guinness Book of World Records* named it the "World's Most Emotionally Responsive Robot." But Cynthia Breazeal knows why. She remembers Kismet's glory days—when those big blue eyes, fuzzy eyebrows, and red rubbery lips reacted to the sound of her voice. It was a time when her robot pal seemed almost alive.

Back in 2000, when Kismet lived in Cynthia's workspace on the ninth floor of the Artificial Intelligence (AI) Lab, meeting the

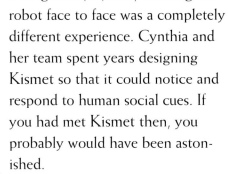

robot face to face was a completely different experience. Cynthia and her team spent years designing Kismet so that it could notice and respond to human social cues. If you had met Kismet then, you probably would have been aston- ished.

In those days, Kismet seemed to know what people were saying. For example, if you walked into the lab and casually asked, "Hey, what's up, Kiz?" the robot would

In April 2003, schoolchildren visited the MIT Museum in honor of the addition of Kismet's head to the museum's collection.

crane its neck in your direction. As you moved closer, its big blue eyes would make eye contact with you and follow your movements. If you said in a sweet, singsong voice, "You are such an adorable robot!" Kismet's face would move closer and smile. But if you scolded Kismet with a stern "Bad robot!" Kismet would pull back in fear.

Of course Kismet could not actually understand English—or any other language, for that matter. Thanks to Cynthia's innovative programming, however, Kismet was able to use pitch and tone to recognize the emotional quality in peoples' voices and respond accordingly. The robot's face could express a variety of "emotions," including happiness, sadness, anger, surprise, disgust, and even exhaustion.

Because Kismet was so expressive, it was sometimes hard to remember that the robot couldn't actually "feel" anything the way humans and animals do. But that was a sign that Cynthia's project was a success. Cynthia was trying to create a robot that could imitate the emotional behavior of a human infant. Her mission was to build a mechanical creature that could use its facial expressions and babbling voice to communicate with humans in a spontaneous, lifelike manner. The people who met Kismet were often astounded that Cynthia had achieved this ambitious goal. At times, the accomplishment amazed Cynthia herself. But the idea of making a robot had been simmering in her head for decades.

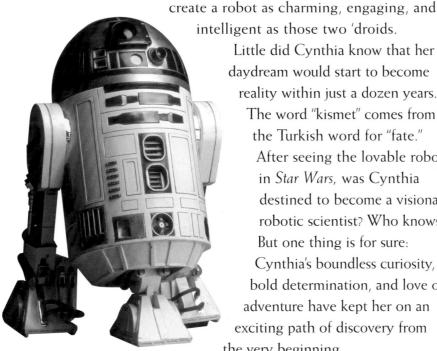

Back in 1977, when Cynthia was ten years old, she watched a thrilling new movie. It was the original *Star Wars*. Like so many other kids, Cynthia was fascinated by the movie's robot heroes, R2-D2 and C-3PO. Cynthia daydreamed that one day she would create a robot as charming, engaging, and intelligent as those two 'droids.

Little did Cynthia know that her daydream would start to become reality within just a dozen years.

The word "kismet" comes from the Turkish word for "fate."

After seeing the lovable robots in *Star Wars*, was Cynthia destined to become a visionary robotic scientist? Who knows? But one thing is for sure: Cynthia's boundless curiosity, bold determination, and love of adventure have kept her on an exciting path of discovery from the very beginning.

When Cynthia saw the original *Star Wars*, she fell in love with the human-like qualities displayed by R2-D2 *(left)*, a spunky, resourceful robot. Cynthia *(above)* chats with Kismet, as it responds to the social cues coming from her voice and movement.

The **speedy**

arrival of little "Cindy"

was a **sign** of things to come

for her parents.

2

SPIRIT OF ADVENTURE

On November 15, 1967, less than an hour after her mother arrived at the hospital in Albuquerque, New Mexico, Cynthia Lynn Breazeal (bruh-ZILL) was born. The speedy arrival of little "Cindy" was a sign of things to come for her parents, Norman and Juliette Breazeal. It turned out their new daughter had quite an adventuresome spirit.

As a toddler, Cindy loved physical challenges. When she was only three years old she spent hours climbing trees with her five-year-old brother, Bill. This daredevil interest frightened her parents so much that Cindy's father sawed off the lower branches of a tall tree in their backyard. He hoped this would discourage the kids from climbing.

Nice try, Dad! Only a few days after the lower branches were removed, Cindy's mother was shocked to find the children high up in the very same tree. Far from being scared, Cindy and Bill were giggling with delight.

Trees weren't the only things three-year-old Cindy loved to scale. One day Cindy's mother looked out her kitchen window to find Cindy strolling behind her brother—on top of a six-foot-tall brick wall! Cindy's attitude seemed to be, "If my brother Bill can do it, so can I!"

At age 2 *(above)*, Cindy loved tackling physical challenges in her backyard with her older brother, Bill *(opposite)*.

On another occasion, Cindy's mom was unloading groceries from the car in her driveway. As she glanced into the backseat to check on her daughter, her pulse quickened: Cindy had disappeared! Juliette frantically started calling out Cindy's name. Seconds later she found her. Cindy was sitting on *top* of the car, smiling broadly at her mother.

As Juliette carefully guided her daughter down, Cindy accidentally cut her foot on the car's windshield wiper blade. The injury turned out to be minor, but Cindy's doctor was concerned. This was the third time the Breazeals had brought Bill or Cindy to his office for stitches in the last few weeks. Juliette realized that she would have to watch her daring children even more closely.

~ California Bound

Cindy's artistic creativity and love of sports were nurtured early on by her devoted parents.

Life was good for the Breazeal family in Albuquerque. Cindy's father, who has a Ph.D. in mathematics, worked at a research lab. Cindy's mother, in addition to caring for her children, was taking courses toward her master's degree in mathematics at the local university. Since both parents shared an interest in science, dinnertime conversations were often sprinkled with lively academic discussions.

Even though they greatly enjoyed Albuquerque, the Breazeals' years there were numbered. In 1971, Cindy's father was transferred to another office in Livermore, California. Although it was a big move, Cindy's parents didn't complain. Going back to California would be a homecoming for Norman and Juliette. They had met and fallen in love when they were students at the University of California, Los Angeles (UCLA), in the early 1960s.

~ Early Life in Livermore

As Cindy settled into her new home, she developed a special interest in animals. She loved her pet goldfish and caught bugs in the backyard just so she could "care" for them. Eventually, Cindy begged for a dog. Her parents, not convinced that she could care for such a large animal, suggested she start with something smaller—like a white rat. Believe it or not, Cindy was delighted with the compromise. She and her brother named their new little pet Miscellaneous Mushmouse. "Mushmouse" came from one of their favorite cartoon characters. As for "Miscellaneous," Cindy thought it would make a funny first name.

Given her obvious love of animals, Cindy's parents were not surprised when she made a special announcement. At the ripe old age of seven, Cindy had decided she would become a veterinarian when she grew up.

Cindy and her brother, Bill, proudly display their first pet, a white rat named Miscellaneous Mushmouse, after a 1960s cartoon country mouse.

~ Grade School Roller Coaster

Cindy's early elementary school experiences in Livermore were wonderful. From preschool through second grade, she thrived and approached her classes with gusto. She loved learning to read and write and enjoyed creating art projects. Her teachers were impressed with her intelligence, creativity, and enthusiasm.

Cindy's blossoming success in school hit a snag as she was about to start third grade. At that time Cindy's parents were investing in real estate and decided to move to another location in Livermore. This meant that Cindy and Bill would have to go to a different elementary school.

Things did not go well for Cindy at the new school. Although she had always lived up to her potential before, her grades were now only average. Cindy's parents worried that she was not being adequately prepared for the state's achievement tests. They also worried that the teachers at the new school were failing to nurture her creative talents.

During her first year at the new school, the TV show *Star Trek* inspired Cindy to write a science fiction story. In her paper, Cindy described an invention that she named the "Z,X,W,7,10,5" machine. The purpose of the machine was to help hungry Klingons find homemade huckleberry pies when they invaded Earth. The machine's "feelings," Cindy explained in the story, were run by a computer.

Cindy's teacher did not respond well to this creative effort. Her only comment was that Cindy should proofread next time.

In addition to indifferent instructors, Cindy and her brother had to deal with racist remarks. Their mother was of Korean ancestry, and Cindy and Bill's distinctive facial features drew unkind comments from other children. It didn't help that Cindy was short for her age. Cindy tried to ignore the insensitive comments, but they often hurt her feelings.

Fortunately for Cindy, her brother came to the rescue. Bill often used humor to soften the blow of cruel remarks. When other kids made fun of Cindy's appearance, Bill would wonder aloud if perhaps they were right. He would exaggerate their criticisms until the comments sounded so ridiculous that they made Cindy laugh. Bill's clever response to the teasing helped his sister learn to take the comments in stride.

To help their children do well in school, Cindy's parents got personally involved with their schoolwork. Every night they looked over Cindy and Bill's assignments and helped them prepare for quizzes. Family field trips also fertilized the children's budding interest in science. The Breazeals traveled to the Exploratorium, a science museum in San Francisco. Here the children played with interactive exhibits and watched amazing science demonstrations.

Cindy's concept of a machine that searches for huckleberry pies reveals her budding talent for creative robot design.

Dinosaur National Park *(left)* and the hands-on Exploratorium *(below)* in San Francisco were two of many field trip destinations for the Breazeal family.

The Breazeal family also visited Dinosaur National Monument, a park that spans parts of Utah and Colorado. Cindy and Bill were fascinated by the hundreds of dinosaur fossils they saw there.

Living in California gave the Breazeals the opportunity to make frequent trips to another wonderful place—Disneyland. Cindy was always amazed at how the amusement park blended the latest technology with popular entertainment. She read books about Disneyland's founder, Walt Disney, and was inspired by his creativity.

~ *Turning Things Around*

Thanks to her parents' efforts at home, Cindy's grades gradually improved. By moving to a different part of Livermore, the Breazeals found better schools for their kids. Even so, by fifth grade Cindy still seemed to be falling short of her potential. She felt discouraged.

Enter fifth-grade teacher Ben Green. Mr. Green approached the art of teaching the way he coached sports: He encouraged his students to throw themselves into their studies, giving nothing less than 100 percent. He frequently had them play high-energy games as a way to practice skills. Often these games involved both physical and intellectual challenges. He also created mini-contests that pitted students against one another, as well as themselves, to solve problems.

Even with Mr. Green's creative teaching methods, Cindy's grades were not as good as they could be. So one day he asked her to

Walt's Wonders

Cynthia Breazeal's fascination with science was nurtured by a theme park in Southern California. On family trips to Disneyland in the 1970s, she enjoyed visiting the part of the park called Tomorrowland, which blended science and fantasy. One exhibit there, "Adventure Thru Inner Space," made audience members feel they had been shrunk so small they could see the world inside a drop of water.

Cynthia read about Disneyland creator Walt Elias Disney (1901–1966), a trailblazer in the entertainment world. Walt Disney produced the first cartoon with sound, *Steamboat Willie,* and the first full-length animated movie, *Snow White and the Seven Dwarfs.* He also pioneered the modern theme park.

When Disney unveiled his grand design for Disneyland in the mid-1950s, most people thought it was doomed from the start. Convinced of his vision, Disney ignored the naysayers. In the early 1950s, for example, artist-designer Claude Coats

was trying to create a rainbow-colored waterfall for a Disneyland ride. A famous engineer viewed a prototype of the waterfall and advised Coats to give up. "In a few days," the engineer warned, "all these separate colors will blend into a drab gray."

When Coats relayed the engineer's gloomy assessment to Disney, Walt replied, "It's kind of fun doing the impossible." Buoyed by Disney's enthusiasm, Claude worked twice as hard to figure out a way to make the rainbow waterfall function. It flowed perfectly for years.

When Disneyland opened to the public on July 17, 1955, problems abounded. Many of the rides did not work properly. Sleeping Beauty Castle sprang a gas leak. The flying elephants on the Dumbo ride were too heavy to be lifted by their machinery. Despite this shaky start, Disney continued to improve his namesake park until it became a success. He had followed an old formula—curiosity, confidence, courage—to craft something new.

meet him after school. Cindy nervously wondered what she had done wrong. Would Mr. Green yell at her?

Far from it. Like a coach consoling a player in a rut, Mr. Green gave Cindy a pep talk. He explained to her that, coincidentally, his wife had been one of Cindy's first-grade teachers. She remembered Cindy as one of the brightest students she had ever had.

Mr. Green urged Cindy to try harder. He explained that if she was willing to put in the effort, she could accomplish great things in school and beyond. Cindy took Mr. Green's advice to heart. She studied longer, asked more questions in class, and revised (and revised) her papers to perfection. Mr. Green was delighted. Cindy's curiosity and enthusiasm were finally back in full force.

As the end of fifth grade approached, however, Cindy was sad to be leaving Mr. Green's class. He had been such a marvelous teacher. Would her next teacher be as inspirational? Cindy then got some great news. Her sixth-grade teacher would be none other than Ben Green. He would be "moving up" with Cindy's class.

Ben Green *(far left)*, one of Cindy's favorite teachers, often applied his coaching talents to the classroom. His high-energy methods motivated students, including Cindy *(front row, third from left)*, to try their best.

Like most
girls her age,
Cindy was interested in

many other things,
from **sports**
to boys.

A Well-Rounded Education

3

Sixth grade with Mr. Green sped by for Cindy. She continued to flourish under his rigorous but supportive teaching style. At home Cindy's parents still played an active role in enriching her studies, spending many hours making sure she mastered the material. Before seventh grade, her parents asked the school to test Cindy for the honors program. The results showed that Cindy was indeed ready for new challenges. Sure enough, she thrived in several honors courses.

But life wasn't all work and no play. Like most girls her age, Cindy was interested in many other things, from sports to boys. She even took a modeling class at a local community college. The class was intended to help young girls explore their own sense of style. It also taught her how to make a positive impression during an interview. Cindy had no idea how that skill would come in handy when she was cast into the media limelight at MIT many years later. But even in her middle school years, learning poise under pressure served her well: Cindy entered—and won—several modeling contests. Not bad for a timid little girl who had once been teased about her looks!

During her middle school years, Cindy *(seen opposite at age 11)* began to learn a lot about herself. Among other interests, she enjoyed and excelled at sports, for which she won many awards *(above)*.

Developing a personal sense of style was one of the skills Cindy worked on in her modeling class.

To continue Cindy's well-rounded exposure, her mother took her to several daylong Women in Science conferences sponsored by the Lawrence Livermore National Laboratory and the Sandia National Laboratories. Here Cindy listened to working women speak about their careers as engineers, computer programmers, and medical researchers.

Although the seven-year-old Cindy had declared she wanted to be a veterinarian, this more mature Cindy realized her career options were still wide open. She wasn't certain exactly what she wanted to do, but she had a hunch it would involve math and science. At the time her mother was working as a computer scientist. Perhaps she would follow in her footsteps? Listening to the advice of working women at the science conferences gave Cindy some new perspectives to think about. It also encouraged her to keep at her studies.

~ Leaps and Bounds

At the same time that Cindy the student was overcoming academic obstacles in middle school, Cindy the athlete was leaping hurdles—literally. Track was one of several sports she enjoyed during her time at Mendenhall Middle School in Livermore. Remarkably fast on her feet, Cindy frequently beat the boys in races. She loved being the first to cross the finish line.

In addition to sprinting in the 50-, 100-, and 220-yard dashes, Cindy developed into an accomplished hurdler. To improve her skill in this area, her father took her to the library to find books on hurdling techniques. He even built hurdles in their backyard so she could practice there. Once again, her hard work was rewarded: She won several first-place ribbons for hurdling.

However, Cindy the athlete was not done. When soccer caught her interest, she approached the sport with customary intensity. In addition to practicing with local teams, Cindy often went to the park with her father to refine her passing, dribbling, and heading skills. Although she loved the action on the field, Cindy was less enthusiastic about the cliques that had started to form among the 13-year-old girls on the team. It was common for one group of girls to pick on a girl who had made a mistake during a game. Because Cindy was a fast, skilled player, none of the kids picked on her. Even so, she found the "cliquey" behavior offensive; why couldn't everyone give up the petty teasing and just focus on soccer? Cindy vowed to stay away from the cliques. Unfortunately, that meant she never developed any close friends on the soccer team.

In eighth grade, Cindy's willingness to try new things led her to a different kind of athletic activity. She and several friends from her honors classes tried out for the cheerleading squad. Cindy was delighted when she made the team. She spent many hours inventing cheers and choreographing moves with the rest of the girls.

At 13, Cindy loved playing soccer but, like many girls, she was bothered by the senseless teasing of weaker players. For her, the soccer clique was not worth joining.

After a while, however, watching sports from the sidelines began to lose its appeal. Cheerleading was "frustrating," she told her parents, because she wanted to take part in the action on the field, not just watch it. After fulfilling her commitment to the cheerleading squad, Cindy retired her pom-poms for good. Even today she finds it hard to share her husband's enthusiasm for watching football or baseball on TV. If she's not actively involved in a sport—out there sweating and pushing herself to the limit—it fails to engage her.

~ Serving Up a Stellar Performance

As Cindy's self-esteem grew, the shy little girl developed into a confident teenager. In fact, just a few months before high school was to start, Cindy set her sights on an ambitious new goal: She took up tennis, hoping to make the high school tennis team.

Most kids who dream of becoming top-ranked tennis players first learn the sport when they're six or seven years old, so Cindy had a lot of catching up to do. On top of that, tennis requires good technique. To become an accomplished player, Cynthia had to hone specific skills. Speed and spirit were not enough. Fortunately, her family had recently joined a local tennis club and her brother Bill enjoyed playing the game. So Cindy had a place to practice and a supportive partner on the court.

As usual, she focused hard on the task at hand. From June to August, she threw herself into learning tennis. Even the scorching summer heat never got in the way of her determination to improve. Cindy soon discovered she had a very good two-handed backhand, but her forehand needed work. Often she would accidentally whack the ball over the fence surrounding the court. To improve her forehand and other tennis strokes, her parents found coaches at the tennis club to help. They also bought an automatic ball machine that could shoot balls at her at various speeds, spin rates, and arc angles.

Cindy's hard work paid off. Not only was she accepted on the Granada High School tennis team—she won the number-one player spot. She held that position through her senior year but never rested on her laurels. She continued to take lessons with coaches and seek out challenging players to compete against.

As a tennis star in high school, Cindy never took her victories for granted. Instead, she recognized the value of continued practice.

16

Cindy approached the sport of tennis with a scientist's mind. With her father's help, she videotaped her swings and analyzed what she did right and which skills needed work. During her high school tennis career, Cindy won many tournaments and was ranked as one of the best players in the area. She even briefly considered becoming a professional tennis player, but the lure of science always won out. Cindy knew she wanted to study medicine or engineering in college. To succeed in these competitive science fields, she knew she would have to focus on her studies, which meant she would have to give her tennis racket a rest.

> Cindy's hard work paid off. Not only was she accepted on the Granada High School tennis team—she won the number-one player spot.

~ Sound Advice

Cindy's parents had always recognized the value of a good education, and they took great efforts to motivate their children. They urged both Cindy and Bill always to aim for the top, particularly with their grades. "If you work for A's," they'd say, "then you'll probably get A's and some B's. But if you set your sights on B's, you'll probably end up with B's and some C's." In short, they said, if a goal is important to you, give it your best effort.

Cindy, at 16, with her brother, Bill, and parents Juliette and Norman Breazeal.

Cindy's parents were sensitive to another issue that was particularly troubling for professional women. Cindy's mother had observed that a number of talented women had never reached a high level of success in their careers because they were almost apologetic about their achievements. Thus, Cindy's parents often told her, "Don't be afraid to toot your own horn."

Exaggerated bragging wasn't appropriate, but they assured Cindy that sharing wonderful news about her recent accomplishments with influential people was a good thing to do. Cindy's parents felt that "spreading the news" could open doors that would help Cindy take her accomplishments to the next level.

~ *Persistence Pays Off*

Commitment and discipline are hard qualities to master, but the rewards are great, as seen here by Cindy's many medals, ribbons, and trophies.

Cindy took her parents' advice to heart, especially when it came to grades. In high school she aimed for A's in all her classes, and she aced physics and chemistry. Unfortunately, she always ended up with B's in math. For some kids this would be considered

respectable work, but Cindy's mathematician parents saw more potential in their daughter. They urged her to keep trying. If she didn't give up, they felt sure, one day Cindy would excel in math.

They were right. From Cindy's junior year on, she received top grades in her math classes. For the first time, she was regularly acing her math tests. Of course, she still had to study hard, but she didn't feel as though she was fighting an uphill battle.

Cindy's persistence with her studies, sports, and other high school interests helped her become a well-rounded young woman. As a result, as her high school graduation approached, Cindy got some exciting news: She had been nominated for the Livermore Boosters Olympian Award. This award was given annually to two seniors (one boy, one girl) who had displayed an "outstanding combination of scholarship, sportsmanship, citizenship, and integrity" throughout their high school years. The winners would receive money for college tuition.

> When she saw all the other girls wearing beautiful dresses, she thought she'd made a big mistake. What had she been thinking?

As part of the application process, Cindy would have to deliver a speech before a panel of judges. What would she talk about? Given her love of science, you might guess she picked a topic in that field. But Cindy chose a completely different approach. She decided to explain how playing on the tennis team had helped prepare her for college. She would talk about how the game had taught her the vital importance of training, persistence, practice, and managing her time. Playing tennis had also helped her learn to recover from setbacks with grace and resilience.

On the day Cindy gave her lecture, her father suggested she wear her tennis outfit for added impact. Cindy was not sure this was such a good idea, but she decided to give it a shot. When she saw all the other girls wearing beautiful dresses, she thought she'd made a big mistake. What had she been thinking?

As it turned out, the risk was well calculated. A few days later Cindy learned she had won the Olympian Award. One of the judges even called her parents to commend Cindy on her moving presentation.

~ A Diploma and a Decision

On June 14, 1985, Cindy graduated from Granada High School in Livermore, California. Her adventures in engineering—and in life away from home—began in the fall.

In addition to winning the Livermore Boosters Olympian Award, Cindy finished her high school career with a high grade point average. Ranked seventh out of 328 graduates in her class, it was clear she had a good chance of getting into a competitive college. But what career would she pursue?

Cindy's childhood dreams of becoming a veterinarian may have faded, but her interest in medicine had not. In hopes of becoming a physician one day, she considered enrolling in a pre-med program. But then, during her senior year in high school, Cindy developed a greater interest in engineering. When one of her teachers heard that she planned to pursue engineering in college, he said, "You do realize that anything you learn in engineering school will soon be outdated, right?" Cindy wasn't worried about her technical knowledge becoming obsolete. "That's why I want to go into engineering," she replied. "I want to be learning and creating all my life."

So, where to study engineering? Cindy first considered joining Bill, who now went by his full name, William, at the California Institute of Technology in Pasadena, where he was studying physics. CalTech, as the school is informally called, is one of the top-ranked science and technology institutions in the United States. But William had reservations about whether it would be right for his sister. Although CalTech had a phenomenal engineering program, William thought the environment might be too male dominated.

Heeding his advice, Cindy and her parents decided to look at some public schools in the University of California (UC) system. They liked the campus of UC, Santa Barbara (UCSB), which had a well-respected engineering program. The school also had a good pre-med program, which Cindy could pursue if engineering didn't turn out to be as satisfying as she expected. So Cindy applied and, happily, was accepted.

One of the administrators of the UCSB engineering program informed Cindy that she would be in good company in her freshman class. The female students in the incoming class of 1985 had outstanding academic credentials—some of the best ever. Cindy also learned that fewer than 5 percent of the incoming engineering students were women. Cindy had competed against boys in a variety of sports in schools, so she wasn't too disturbed by this lopsided ratio. She saw this next step as an exciting opportunity to get a top-notch education on one of the most beautiful campuses in the nation. What would it be like to go to school away from home? Cindy couldn't wait to find out.

Cindy was thrilled
at the idea of *living*
and *learning*

on such a
beautiful campus.

ENGINEERING A FUTURE

4

L ike most big brothers, William gave Cindy advice as she started her college career. His most important warning was, "Fall behind in your homework and you're doomed!"

William knew this from experience. During his first two years at CalTech, he had learned the vital importance of keeping up-to-date with assignments. He knew that the science concepts at UCSB would be taught sequentially, with key ideas building on each other. Students who didn't master the material bit by bit were sunk. He told Cindy to prepare for new information to come at her at a mind-boggling rate. "It'll be like trying to drink water from a fire hydrant," he joked.

During a summer orientation, the engineering school staff offered similar words of advice. They encouraged the freshmen to befriend each other and steer clear of students who took their partying more seriously than their studying. Unfortunately, that might not be easy: In the mid-1980s, UCSB was ranked by a major magazine as one of the nation's Top 10 Party Schools.

Cindy was thrilled at the idea of living and learning on such a beautiful campus. UCSB is located on a cliff overlooking the Pacific Ocean. Some of the dorms are a short distance from the beach. For some people the beach might have posed a distraction.

The UCSB campus takes up nearly 1,000 acres on the California coast. The school's fantastic scenery can be just what a student needs after a tough week of studying. Above, Cynthia takes a break to do some rock climbing.

At Anacapa Dormitory *(top)*, Cindy met serious-minded students who, like her, sought a healthy balance between work and fun. A Hawaiian theme party offered a welcome break for Cynthia and a friend *(above)*.

But Cindy knew why she was there—to get a solid education. The key would be finding an environment that supported serious studying. That environment turned out to prevail at Anacapa Dormitory's Hall of Academic Pursuit—otherwise known as Nerdy Hall.

Cindy had heard about Nerdy Hall from a girl she met in a bookstore during her summer visit to the campus. The girl highly recommended two sections of Anacapa dorm, which were reserved for students who were dedicated to studying. In the evenings, students were generally quiet and considerate of others. No blaring stereos, no impromptu pizza parties, just hard-core studying. The girl in the bookstore added that Nerdy Hall had a great support group, with older students helping younger ones.

When Cindy arrived at the UCSB campus in the fall of 1985, she was delighted to discover that the students in her dormitory were as dedicated to fun as they were to their studies. Although they kept their noses in their books Monday through Friday, they made a point of having a good time on the weekends. They played flag football and went to basketball games and movies. When a group of students gathered to watch football on TV, they yelled their lungs out. Anacapa was a wonderful place to live and study.

When Cindy needed a break from her studies, she sometimes played tennis. She didn't play the intensely competitive tennis of her high school career but pickup matches just for fun. For someone of her competitive nature, fun could be difficult. However, Cindy tried to relax and take her father's advice. When she was packing

24

her racket for college, he had made a point of saying she needn't worry about playing up to her usual standards. "In fact," Cindy's father had joked, "it's all right to let the boys win once in a while!"

~ Fun with Physics

Like all freshmen, Cindy had to take courses in English and math. All UCSB engineering majors were also required to take a year and a half of introductory physics. Thanks to her professor, Roger Freedman, physics turned out to be one of Cindy's favorite classes.

A Van de Graaff generator is designed to create static electricity for the purposes of experimentation. For the non-scientist, it can be a hair-raising experience.

Professor Freedman brought the concepts of physics to life with dramatic demonstrations. Once he used a children's red wagon to teach the students about Sir Isaac Newton. Newton was a famous 17th-century mathematician and scientist who figured out laws about how objects move in the physical world. To demonstrate the concepts of force and motion, Professor Freedman sat in the wagon, which was equipped with a fire extinguisher. Then he squeezed the handle on the extinguisher. The resulting *WHOOSH!* propelled both the wagon and Professor Freedman across the room at more than 20 miles per hour.

Another time Professor Freedman brought in a machine called a Van de Graaff generator to show the effects of static electricity. When students touched the generator, their hair stuck out in every direction. Professor Freedman explained that because each strand of hair carries the same type of electric charge, it repels all the strands near it.

Cindy got a kick out of Professor Freedman's demonstrations. She worked hard to master the concepts. To make sure they really understood the material, Cindy and her friends tackled textbook problems that hadn't been assigned for homework.

When they were stuck, they visited Professor Freedman during his office hours with questions, eager to learn as much as they could about the world of physics.

~ Real-World Experience

At the end of her sophomore year, Cindy was hungry for some practical engineering experience. Answering test questions about electrical circuits and electromagnetism was one thing, but trying out this knowledge on the job was a different matter. Electrical engineers design and build new products, such as cell phones and computer processors. Cindy was eager to give it a try.

Her extra efforts were appreciated—and rewarded. When she received her first paycheck, it was for much more money than she expected.

On the lookout for opportunities, Cindy heard about a summer internship opportunity at Xerox, one of the world's largest technology companies. The internship was offered at an office in El Segundo, California—not too far from UCSB. Cindy applied for the position and was accepted.

Before her first day at Xerox, Cindy's parents offered her some work advice. Although they knew their daughter would be a conscientious employee, they reminded her about the importance of making a strong impression. They advised her to get to work early and leave later than others.

Following her parents' suggestion, Cindy often worked more hours than required. Although the long days were exhausting, Cindy found that her parents were right: Her extra efforts were appreciated—and rewarded. When she received her first paycheck, it was for much more money than she expected. Had someone made a mistake? Far from it; one of Cindy's supervisors had adjusted her timesheet to reflect the extra hours she had worked.

Part of Cindy's job at Xerox involved microchip design. Computer microchips are tiny machines that process information, make calculations, and control the flow of information. Before Xerox could produce a particular kind of chip in large quantities,

Microchips, like the one below, are found in countless electronic devices, from computers and calculators to digital cameras, TVs, and DVD players. Cynthia spent one summer testing microchips at this Xerox site *(left)*.

it had to make sure the chip worked perfectly.

One day Cindy was working on a chip that had passed all the standard tests. But she still wasn't satisfied; something didn't feel quite right. So she ran some additional tests, and sure enough the chip was flawed; it didn't work properly under certain conditions. Her grateful boss thanked Cindy for catching this goof. Had this defective chip been sent to production, Xerox would have wasted a great deal of money and effort.

~ Deciding on Grad School

As an upperclassman at UCSB, Cindy knew it was time to start thinking about what she would do after graduation. For a while, she considered graduate programs that would prepare her to become an astronaut with the National Aeronautics and Space Administration (NASA). Professor Freedman had sparked her interest in astronomy, and she was intrigued by the idea of exploring the mysteries of the universe.

However, Cindy's career plans changed during her senior year when she listened to a friend discuss plans to build robotic planetary rovers for NASA. Assembling unmanned machines that could explore other planets fired Cindy's imagination. She remembered her fascination with the robotic characters in the *Star Wars* movies. Was it really possible to learn how to build robots for a living?

Cindy looked at graduate schools with a new eye. She applied to 14 different universities. Then she waited nervously for a reply. Despite having an impressive 3.8 average and graduating magna cum laude (with high honors) from UCSB, Cindy did not take her grad school acceptance for granted. She was applying to some of the most competitive graduate programs in the country, so she knew there was no guarantee she would get in.

Cindy needn't have worried. Nearly every school accepted her and offered her a teaching fellowship that would provide her with financial support. Her top choice, the Massachusetts Institute of Technology (MIT), offered her a full fellowship that included tuition and a $13,000 stipend, or academic allowance. Cindy was delighted.

After accepting MIT's offer, Cindy took another important step. She shopped around for a doctoral advisor. In graduate school,

In May 1989, Cindy graduated with high honors from UCSB with a degree in electrical and computer engineering *(above)*. Next stop: the AI Lab at MIT. Two lovable Hollywood stars *(right)* serve as enduring and endearing AI inspirations for Cindy.

a doctoral advisor is a professor who serves as a mentor, helping students develop academically and also funding their research. To find just the right person, Cindy read research papers written by several professors and eventually contacted a man named Rodney Brooks. Professor Brooks worked at MIT's Artificial Intelligence (AI) Laboratory. (Today the lab is part of MIT's Computer Science and Artificial Intelligence Laboratory, known as CSAIL.) AI is a branch of computer science devoted to creating programs that empower machines to act in ways that mimic human or animal intelligence.

Professor Brooks was a very well known and respected expert in the study of autonomous robots. Autonomous robots are machines that are programmed to function independently and make decisions on their own. They sense and gather information about their environment and process it to perform tasks such as picking up objects or navigating around obstacles. Autonomous robots are different from other types of machines run by software programs because the robots don't follow a set of scripted commands, like "move left, then move right, and then pick up the rock." They can move through an unpredictable environment and make "intelligent" decisions based on what their sensors pick up.

Cindy met with Professor Brooks and was immediately taken with his enthusiasm, warmth, and irreverent sense of humor. She was confident that he was just the type of mentor who would inspire and challenge her as she began to explore the fascinating world of robotics.

What has six legs,
19 motors,
and lots of little computers

and scampers
over tough terrain?

5

THE MOBOT LAB

I n 1990, Cindy—or Cynthia as some people now called her—
arrived at the MIT campus in Cambridge, Massachusetts. She
had come to study artificial intelligence, but she was surrounded
by many examples of the real thing. The students and professors
she encountered were among the smartest, most disciplined,
and most creative people she had ever met. She was immediately
impressed with the environment of camaraderie, a friendly
fellowship and willingness to pool resources to solve problems.
She knew that, if she worked hard and made the most of every
research opportunity, she would eventually make an important
contribution to the field of AI.

At MIT's Mobot Lab,
Cynthia *(opposite)*
poses with Hannibal
(red) and Attila (gold),
the two six-legged
robots that she
programmed to
interact with
unpredictable
environments. This
Earth photo *(above)*
was part of a "set"
designed to test how
robots like Attila and
Hannibal might walk
on the Moon.

~ A Fearless Leader

Cynthia's advisor, Professor Rodney Brooks, directed and supported
most of her graduate student opportunities. He had been working
in robotics for some time and had already contributed breakthrough
ideas to the field. For example, in the mid-1980s, Professor Brooks
had begun thinking about how things move, specifically insects.
He knew that insects could travel more efficiently than any mobile
robot that existed. Their tiny size aside, ants, for example, can

navigate more types of terrain than humans or machines. Professor Brooks wondered: *Could he create a mechanical ant that would make the same maneuvers as a real one?*

At first it seemed as though such a robot would need an enormously complex computerized brain to handle all the computations necessary to sense its surroundings and move its limbs appropriately. Then Professor Brooks hit on a radical idea: *What if I made the robot's behavior as simple as possible?* The key would be to give the robot lots of sensors to learn about its environment, then program the machine to respond automatically. This would free the robot from weighing the pros and cons of each movement. Instead of a single big computer to make all the calculations, the robot could have many smaller processors that could run at the same time and communicate with each other.

Staying true to his ideal of robotic simplicity, Professor Brooks realized that his automated insects didn't need to walk perfectly. After watching videos of real insects traveling over rough terrain, he noticed that they often stumbled but were able to quickly correct their balance. These ideas led to the creation of Genghis, the AI Lab's first robotic insect. It was named after a famous 12th-century Mongolian military leader, Genghis Khan.

> Professor Brooks wondered: Could he create a mechanical ant that would make the same maneuvers as a real one?

While Professor Brooks was working on Genghis, NASA scientists were trying to design a robot that could eventually explore the planet Mars. But the rover model NASA was working on in the late 1980s weighed more than one ton and would cost about $12 billion to build. When Professor Brooks read about this robot, he thought of his own work. *Did NASA really have to spend that much money,* he wondered, *for a single robot that couldn't even move very fast?*

Inspired by his success with Genghis, Professor Brooks and his colleague Anita Flynn wrote a paper suggesting an alternative approach. Instead of sending one 2,200-pound robot to Mars, why not send a hundred 2.2-pound robots? This approach offered several advantages. For one, it would take less time to develop these rovers. In addition, some of them could be risked for exploring dangerous locations and they would not have to be as carefully supervised. These small robots could also cooperate in an exploration mission without having to communicate with one another. The paper was published in 1989 in the *Journal of the British Interplanetary Society* with the whimsical title, "Fast, Cheap, and Out of Control: A Robot Invasion of the Solar System."

In the early 1990s, Professor Brooks secured funding that would enable his team to build small micro-rovers to help answer the question of whether insectlike machines could one day explore planet surfaces more safely and less expensively than humans.

The Mobot Lab (short for "mobile robot laboratory") would be a dream assignment for any new graduate student. The timing of Cynthia's arrival at MIT couldn't have been more perfect.

~ Making Mobots

What has six legs, 19 motors, and lots of little computers and scampers over tough terrain? There are actually two answers to this robotic riddle: Attila and Hannibal. These were the twin robots that Cynthia helped create at MIT.

Attila and Hannibal were identical except for their color. Attila was gold; Hannibal was red. Made mostly of metal, each micro-rover had six legs. Three actuators (motors) were attached to each leg, so the leg could move in different directions. One actuator moved the leg back and forth. Another moved it up and down. And the third behaved like an elbow, bending the leg at a midjoint. When robot builders brag to each other, they often talk about "degrees of freedom," which has to do with how many different ways an

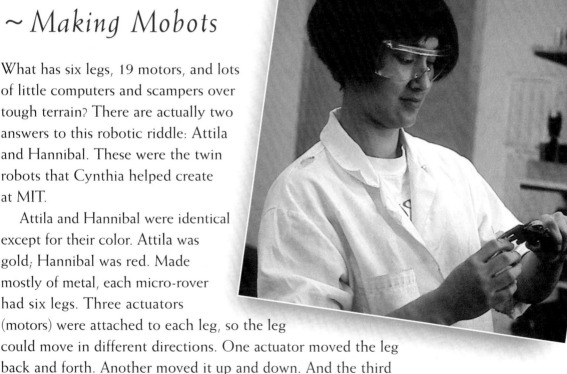

At the Mobot Lab's Machine Shop, Cynthia measures one of her tools by using calipers— an instrument that measures the diameter or thickness of an object.

object can move. This depends on the structure of an object. A leg that can move along three pathways is said to have three degrees of freedom. Each of Attila and Hannibal's bodies had more than 19 degrees of freedom. (The 19th degree of freedom related to the movement of the robot's spine to which the six legs were attached.) This was an impressive range for such small robots.

~ Sensitive by Design

To enable the robots to explore their environment and avoid obstacles, Cynthia and another graduate student, Colin Angle (the student who developed Genghis with Professor Brooks), gave Attila and Hannibal more than 60 sensors each. Both robots' heads were given two touch sensors that resembled whiskers.

Hannibal and Attila are designed to respond to hazards, navigate rough terrain, and detect and deal with failures. To test the robots in a Moon-like environment, the lab created a "sandbox," where robots try to avoid obstacles while traveling on gravel and sand.

If these whiskers bumped into something, they sent a message to the legs to move back. Each leg was given force sensors that let the robot know if its movements were being blocked. Inside each robot was a sensor that would let the robot know the angle of its body. This helped the robot maintain its balance. In addition to their many sensors, Attila and Hannibal were equipped with camera systems that enabled them to identify people and follow them around.

~ A Cure for Breakdowns

As Cynthia and her team designed the robots, they developed the computer software that would help the robots operate. Each body part—the six legs, the body, and the head—had its own microprocessor. The software would analyze the data from the many sensors and quickly make sense of it so the motors could respond by moving appropriately. In addition, the microprocessors needed to be able to share information. The goals were to make the robot insects perform a variety of behaviors, including standing up and walking over obstacles.

But it wasn't easy. Machines can break down, and Attila and Hannibal were no exceptions. There were frequent mechanical and electrical problems to deal with, like a malfunctioning sensor or a stuck gear. What if this happened while the robot was on the Moon? Cynthia was intrigued by the idea of writing software that could help the robots figure out ways around breakdowns.

Taking on these technical challenges required a lot of thinking, but it also made every day in the lab fun and interesting because the team had to be creative. What also made the work exciting

Humor never hurts in the lab. Cynthia created a team T-shirt after every new robot was built. Here, an alien family picnicking on the Moon is being invaded by robot ants.

was knowing that solving these problems would one day benefit the entire AI research community. Part of doing good research is sharing what you've learned with other scientists, such as through journal papers and conferences. That's how science progresses.

Eventually, Cynthia came up with a solution designed to reduce the robot breakdowns. She wrote software that let the robots work around failures of their sensors or other mechanical parts. The software told the micro-rover to identify when its parts were faulty, then figure out how to function with the remaining parts. The challenge was to help the robots respond in real time. In addition to addressing breakdowns, Cynthia developed models—ways to describe a process or system—for how a six-legged creature should travel over rough terrain. She based her models on the behavior of real six-legged insects.

Cynthia published her research results in a 150-page thesis that she wrote for her master's degree, titled "Robust Agent Control of an Autonomous Robot with Many Sensors and Actuators."

At the National Air and Space Museum, Cynthia met former astronaut, Senator John Glenn. In 1962, he became the first American to orbit Earth.

~ The Mobots Meet Their Public

Not surprisingly, Hannibal and Attila received a lot of attention. The Smithsonian Institution's Air and Space Museum invited Professor Brooks and his team to display Attila in a planetary rover exhibition. Cynthia was thrilled to visit Washington, D.C. While she was there, she got to meet a former astronaut, Senator John Glenn.

On another occasion the Planetary Society, the world's largest space interest group, held a multinational event for planetary rovers in Death Valley, California. Cynthia attended with Hannibal. The terrain in Death Valley is quite similar to the kind a robot might encounter on Mars. So Cynthia got a chance to see how Hannibal behaved in a more authentic environment. Hannibal even took a turn at "docking" with a Soviet-built rover by walking up a ramp and onto its back.

With the success of Hannibal and Attila, Professor Brooks and his team were ready to move on to even bigger projects.

~ Honoring HAL

On January 12, 1992, Rodney Brooks invited Cynthia and his other graduate students to his house for a birthday party. The guest of honor was HAL, the fictional computer in the famous science fiction movie 2001: A Space Odyssey, directed by Stanley Kubrick. According to the 1968 film, January 12, 1992 was the date when HAL had been switched on. To honor the real date's arrival, Professor Brooks staged a birthday party for the fictional HAL.

What could Professor Brooks's team do over the next 10 years to make the fantasy of HAL a reality in their lab?

2001: A Space Odyssey was an important movie to Professor Brooks. He had first seen it when he was a gadget-building teenager in Australia. The young Brooks was most impressed with HAL, the remarkable computer that used vision and speech to communicate intelligently with human characters. In fact, it was HAL that had inspired Brooks to study artificial intelligence.

During the party, the students discussed the sharp contrast between the robotic insects they were building and the intelligent machines that HAL represented. The year 2001 was still almost a decade away. What could Professor Brooks's team do over the next 10 years to make the fantasy of HAL a reality in their lab? She did not know it at the time, but the person who would eventually "deliver on the promise of HAL" (as Rodney put it) was none other than Cynthia.

~ A Radical's Sabbatical

In 1992, Professor Brooks took a sabbatical from MIT. A sabbatical is a break of several months or a full year that professors can use to travel, teach at another institution, or explore new topics. Cynthia would miss her mentor's guidance, but she knew this would be a great time to work independently.

During his sabbatical, Professor Brooks had a chance to think about his insect robots. As impressive as they were, they were nothing like HAL. Originally Professor Brooks had thought that when he returned from his sabbatical, his team would build robotic reptiles, then perhaps artificial mammals, and then a synthetic person. But realizing that life is short, Professor Brooks decided to go for broke. Despite all the technical obstacles, he made up his mind that his team would attempt to create an autonomous robot in human form. Cynthia and his other students, he knew, would be ready for the challenge.

~ A Cog Is Born

Started in 1993, that dream of a humanoid robot eventually became Cog. The new robot was about the size of a human body from the waist up and had heavy metal arms. Unlike Attila and Hannibal, Cog lacked legs. Professor Brooks, Cynthia, and others thought they should focus their efforts on improving the robot's vision and hearing rather than its mobility. Cynthia came up with the name "Cog." It referred to the robot's mechanical side—like a "cog," or tooth, in the rim of a gear. And because "cog" is an abbreviation for cognition or awareness, it also referred to its intellectual goals.

> Cynthia would miss her mentor's guidance, but she knew this would be a great time to work independently.

The AI team's ambitious mission was to make Cog seem alive, intelligent, and capable of interacting with humans. To achieve its

38

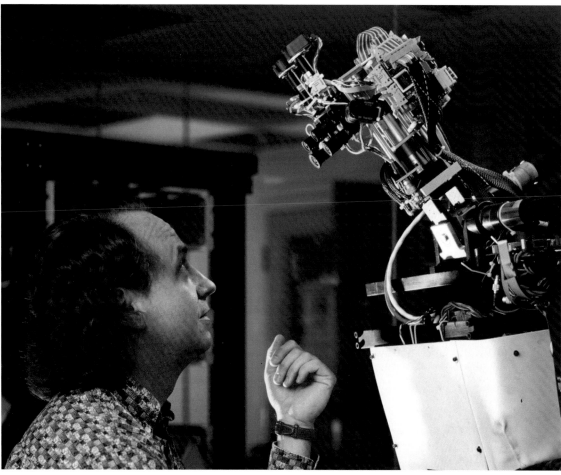

cognitive goals, the robot would have to quickly "see" and analyze its environment, then respond by moving smoothly. Anyone looking at Cog would know it was a machine. Once people interacted with it, though, the team hoped they would be amazed at how well the machine could spontaneously mimic human behavior. Any jerky movements would destroy that illusion.

Rodney Brooks marvels at Cog. Some examples of human behavior that Cog models include turning its head to follow a moving object and playing with a Slinky.

~ The Eyes Have It

The AI team had many energetic debates about how to make Cog seem alive. Giving the robot a sophisticated vision system, they knew, would be essential. In human conversation, eye contact is one of the social cues that tells a person when to speak and when to listen. Job seekers are often told that one of the keys to success

Recipe for a Robot

Robots come in all shapes and sizes. Some have heads; some don't. Some robots walk on two legs, others on six, and still others travel on wheels. As varied as robots are in appearance, so are they in function. Some robots build automobiles, mow lawns, or vacuum floors; others will one day assist astronauts in space. Many "living machines" today are produced strictly for entertainment. Despite their diversity, all robots share the following basic elements:

First, a robot's body is connected by a variety of clever mechanisms that enable it to move.

Second, a robot has sensors that collect information about the surrounding environment. This can include sights, sounds, temperature, and movement.

Third, a robot contains motors, often called actuators. Actuators work like muscles and can provide a robot with mobility. Actuators can also increase the number of "degrees of freedom" a robot has. If a robot's head can turn left and right, that's one degree of freedom. If its head can also move up and down, the head has two degrees of freedom. The more degrees of freedom a robot has, the more realistic its movements will appear.

Fourth, in order to operate, a robot requires a power source. Most robots have an onboard battery or can be plugged into an electrical wall socket.

Finally, a robot needs computer programs, or software, to process information and make decisions. These programs include many algorithms—logical step-by-step procedures for solving problems. Algorithms help the robot "think," act, and learn. The software also enables a robot to make its own spontaneous decisions based on sensor information.

How these five elements are developed depends on one key question in robot design: What is the machine's purpose? Most of Cynthia's robots are designed to interact with people and display social intelligence, so the elements of her robots reflect these and other goals.

Shown at left, Rover, the "sheepdog robot," shows how robots can differ, even with the same basic design elements. Rover is the first-ever autonomous robot designed to control the behavior of animals. It can safely gather and herd flocks of ducks (the same animals that shepherds often use to train their sheep-dogs).

ROBO WORLD

during an interview is to make eye contact with the interviewer. The AI team therefore wanted Cog to identify faces and to maintain eye contact with them. To locate a person's eyes, Cog would need two kinds of "eyes" itself. One type would provide a wide-angle view of the world; the other would offer close-ups.

Cynthia helped build Cog. She spent hours figuring out just what materials to use, then many more hours piecing them together. Her biggest task, though, was to design Cog's first visual behavior system—how it would behave in response to something it "saw." If someone entered the room, Cog would turn in that person's direction and track them as they moved around.

Although the development process was stressful at times, Cynthia and the other team members managed to keep their sense of humor. As news of Cog spread, people began visiting the MIT Web site for more information. The graduate students were responsible for maintaining the Frequently Asked Questions section. One of the questions in the section was this: "Are you worried that your robot might get too intelligent or too powerful?" The grad students' irreverent response: "No—we have programmed the robot to spare our lives in the event that it ever attempts to organize its brethren in a bloody revolution against the human race."

Rodney Brooks once said, "The act of creating a thinking robot forces us to ask the right questions [about] how intelligence works."

~ More Than Meets the Eye

Visitors to the AI Lab thought Cog's long metal arms looked intimidating. Once they saw those arms in action, however, they were reassured: Cog could play a drum and even play with a Slinky toy.

One day, during a videotaping, Cynthia held up a whiteboard eraser and shook it. Cog moved toward the eraser and touched it with its arm. Cynthia waited a moment, then shook the eraser again. Once more, Cog reached out for it. These actions seemed like no big deal at the time, but later, when the team members viewed the tapes, they were quite surprised. It appeared that Cog was playing a turn-taking game with Cynthia—a more sophisticated task than the robot had originally been programmed to do.

This discovery thrilled Cynthia. As she watched the video of her interactions with Cog, she thought how babies learn about the world from their parents. Even though they know infants can't understand words or speak back, mothers and fathers talk to their babies almost immediately after birth. Parents sometimes pause after speaking, hoping the child will do something in response. Little by little, over a period of years, the baby learns how and when to communicate with words.

Norman and Juliette Breazeal proudly celebrate Cynthia earning her master's degree from MIT in 1993.

Could Cynthia create a robot that could socially interact with people? It would look like a machine, of course, but its facial expressions and "senses" would appear like those of a living creature. Cynthia imagined that she could program this robotic baby to display many of the same social cues that human babies use to engage their parents. It would make eye contact. It would use its face and voice to let others know when it was happy, sad, or surprised. She would also make the robot look youthful, in hopes that people would respond to it in a nurturing or playful way, much like a caregiver would with an infant.

The more Cynthia thought about this robot project, the more excited she became. Outside science fiction, no one had yet gone in this new direction. People interact with each other all the time using their facial expressions to communicate. But to have humans and robots interact with each other using expressions associated with emotions—now that was new territory no one had explored.

Here, at last, was a research project whose idea she could call her own. So with help from Professor Brooks and the other graduate students, Cynthia set out to create a younger sibling for Cog.

How could
she make the
robot's **face**

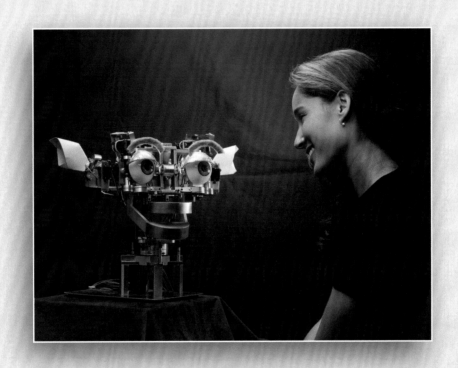

convey the spectrum
of human **emotions?**

A HEAD OF ITS TIME

6

By 1997 Cynthia had been at MIT building robots in Professor Rodney Brooks's lab for seven years. Even though she had played a key role in designing and programming Attila, Hannibal, and Cog, these robots explored different questions. Her new project was to create an expressive robot that socialized with people. It seemed a daunting job. How could she make the robot's face convey the wide range of human emotions? As English biologist Charles Darwin had noted in the 19th century, humans have the most complex and versatile facial expressions of any animal species.

A close-up of Cynthia using a caliper *(above)*. Realizing that Cog would not allow her to explore her new research questions, Cynthia turned to building a new robot—Kismet *(opposite)*.

As Cynthia thought about the challenges ahead, she turned to an inventor from the past whom she had admired since the sixth grade—Leonardo da Vinci. What would the artist-scientist who was known for thinking beyond the limits of his time do in her situation? Cynthia guessed he would think deeply about the central questions, marshal the necessary resources, and forge ahead.

So that's what she did.

To create a robot that could imitate infant behavior, Cynthia began by reading books and articles about developmental psychology. This branch of psychology focuses on how children learn, grow, and behave. Babies are born with instincts that help them survive and communicate with their mother and other adults. Cynthia began to think about what instincts, or motivational drives, her robot would need. As their brains develop, babies depend on input from the five senses, especially sight and sound.

> As impressive as Cog was as a robot, its size tended to intimidate strangers.

Cynthia wondered how she might improve on Cog's ability to see and hear.

During this early phase, Cynthia also searched for a unique robot name. She wanted something that was playful but neither traditionally male nor female. She loved "Gizmo," but Steven Spielberg had already used that for a character in his 1984 movie *Gremlins.* One day the new name hit her: She would call her robot "Kismet," another word for "fate."

When Professor Brooks first heard about Kismet, he was confident that Cynthia's research would break new ground. Kismet was wildly ambitious, intellectually daring, and conceptually creative. These were qualities that Professor Brooks admired and tried to instill in all of his students.

~ What Big Eyes You Have!

As impressive as Cog was as a robot, its size tended to intimidate strangers. Cynthia was therefore determined to make Kismet look both youthful and irresistibly charming. Her goal was for people to want to interact with the robot socially. If she wanted people to treat Kismet like a baby, Cynthia knew she'd have to make her robot much smaller than Cog. But if she made Kismet the size of an actual infant, it wouldn't have room for all its gears and wires. Plus, getting a robot to have an expressive face would be tricky enough without adding working arms and legs.

Unlike Attila and Hannibal, Kismet would not need to travel. So what if this robot was just a mechanized head? To foster the infant-caregiver feel, Cynthia mounted the head on a low table, bringing seated visitors eye to eye with the robot. She also added a movable neck. This enabled Kismet to move its head closer to things it "liked" and pull away from things that "frightened" it. After reading about what makes adult humans bond with babies, Cynthia decided to give Kismet unusually large eyes. She hoped these cartoonlike "peepers" would draw people in and make them want to communicate with the robot. To make Kismet's eyes more expressive, Cynthia gave them large lids, long eyelashes, and bushy eyebrows. To make Kismet's eyes "see," Cynthia and her team adapted Cog's visual system. Four cameras were placed on Kismet's face—one inside each blue eye, one between them, and one where Kismet's nose would be (if it had one). Two of the cameras gathered a wide-angle view of the room; the other two captured close-ups.

To encourage people to interact with Kismet, Cynthia installed two of the robot's four cameras inside its big blue eyes.

As she designed Kismet, Cynthia wanted to make it clear that the robot's head was not human. Much of the research she read stated that people are put off by robots that resemble the human body too closely. She therefore made no effort to hide Kismet's metal bars, gears, and wires. In addition, she gave Kismet a pair of cute pink paper ears, resembling those of a pig. For Kismet's lips, Cynthia discovered that surgical tubing provided excellent flexibility and could easily be colored with a red pen.

The next question was an important one: How could she enable Kismet to physically express, or show, its internal "emotional" state? To find the answer, Cynthia consulted resources on human psychology; she also researched techniques of classical animation. When endowing a cartoon character with lifelike emotions, she learned, you should consider the following factors:

- Keep it simple—try for only one emotion at a time.
- Make the motions and transitions as fluid as possible.
- Believability is more important than realism.
- Don't worry about precise lip-synching. Cartoon lips that are too accurate look unnatural.

Keeping these tips in mind, Cynthia was ready to get down to the mechanics of Kismet.

Kismet's facial expressions reveal its "emotional state." The five human emotions it is expressing here are *(clockwise from top):* surprise, tiredness, happiness, sadness, and anger.

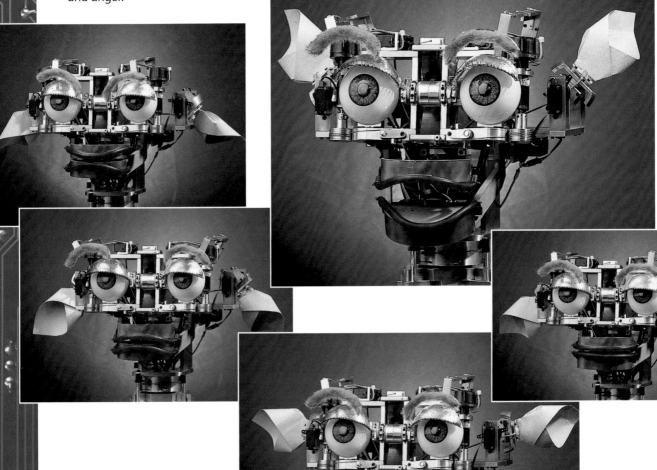

Putting a Human Face on Robotics

Kismet is not the only robotic head around. Take K-Bot, for example. Built in 2002 by David Hanson, a doctoral student at the University of Texas at Dallas, K-Bot's insides look a lot like Cynthia Breazeal's most famous robot, Kismet. K-Bot's interior contains digital cameras, electrical wires, tiny motors, and microprocessors. But unlike Kismet's exterior look, K-Bot's visual design follows a different direction.

David's robot is covered with a soft fleshlike plastic; except for the fact that it has no hair, it resembles a human face. But not just any face: K-Bot was modeled on Hanson's lab assistant, Kristen Nelson—hence the "K" in its name. Low on cash, David built K-Bot for about $400, using parts from hobby, crafts, and hardware stores.

When David hooks K-Bot to his laptop computer, its face moves. It smiles one minute and frowns the next. David spent countless hours studying the anatomy of human facial muscles to develop the movements of his creation. That's because it is incredibly challenging to imitate human facial expressions. When we smile or frown, there are dozens of muscles, small and large, making it happen.

Another challenge to creating a human-faced robot is a psychological one. Many roboticists, including Cynthia, support roboticist Masahiro Mori's theory of the "uncanny valley." According to this theory, people respond positively to artificial creatures that look human—but only up to a point. If the creature looks too much like a human, people will expect it to act like a human. If it doesn't, people will feel uncomfortable interacting with it.

When Cynthia met David in 2002, she explained to him why she does not pursue the use of realistic human faces in her own work. As of 2005, no roboticist has figured out how to create robot faces that move realistically enough. She says, "The focus of my work is social interaction with robots. If my robots look creepy, people will be turned off and not want to communicate or collaborate with them."

K-Bot, seen here in frontal and profile views, has 24 artificial muscles that allow it to make 28 facial movements. The robotic head also has cameras in its eyes to identify and respond to people.

~ Building Kismet's Brain

Once Kismet's look was complete, it was time to put the robot into action. But assembling Kismet's sensors and actuators and writing its software were monumental challenges. For the robot to gather information and respond in real time, it needed a system of "mini brains" that could communicate with one another via networked computers.

Cynthia's enthusiasm was contagious, and her team-building skills were well honed. She was able to pull many talented individuals together to cooperate on Kismet.

From the beginning, Cynthia realized she would never be able to do everything herself. To end up with a viable robot, she would have to ask for help. She had to figure out which tasks she could handle on her own and which required input from others.

Fortunately, Cynthia's passion for the project inspired many graduate students in the AI Lab to join in her effort. Cynthia's enthusiasm was contagious, and her team-building skills were well honed. She was able to pull many talented individuals together to cooperate on Kismet. She began to see how playing on so many sports teams all her life was having a positive impact on her science career.

~ A Hodgepodge of Computers

Like the system that ran Hannibal and Cog, no single computer controlled Kismet. Instead, 15 independent computers communicated with one another through a complex network. To make matters more challenging, the network had to connect computers that were run by different operating systems. An operating system is a master computer program that stores and organizes a hard drive's files, runs all the software, and controls the keyboard, mouse, and other peripherals. All 15 computers were required to get Kismet up and running.

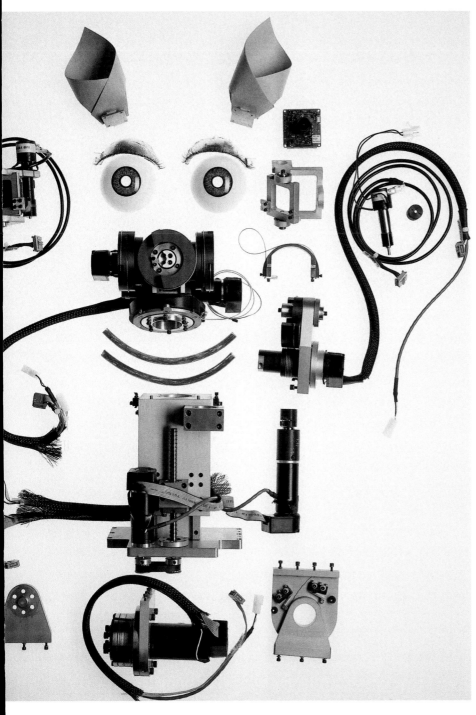

This "deconstruction" collage of Kismet's parts appeared in a book called *Robo Sapiens,* in which Cynthia's ground-breaking research is discussed.

Nine of the 15 computers operated Kismet's vision system. They were programmed to identify the types of things that a human infant enjoys seeing: bright colors (baby toys), objects with skin tone (people), and the movement of nearby objects. Kismet was also capable of eye contact and had lids that could blink. When it was time for Kismet to respond to someone, it would raise its eyebrows and break off eye contact, just like humans do.

Kismet's auditory system was also complex. Its programs analyzed the pitch and intensity of whatever a person said into a microphone. Without understanding specific words, Kismet could determine if it was being praised or scolded and respond accordingly. One of Kismet's programmed responses was to "talk." Kismet's synthesized speech was designed to sound like an infant's babbling. To create the feeling of a true conversation, Kismet filled in awkward silences with its babble.

~ *Kismet's Drives*

As any parent knows, infants are needy. When hungry or thirsty, they cry. If they feel scared or lonely, they cry. If their diaper is wet, they cry. Evolution has provided babies with a powerful means of notifying caregivers, "Hey, how about a little help over here?"

Robots don't eat or drink. They don't need hugs. They don't wear diapers. So how could Cynthia make Kismet act like a needy infant? Kismet's sensors collected detailed data about nearby sights and sounds, but what should Kismet do with this information? How should it respond?

After much thought, Cynthia figured out that Kismet's reactions to its environment would come from three programmed "drives," or "needs." One would be a social drive, or the desire to be stimulated

> **Robots don't eat or drink. They don't need hugs. They don't wear diapers. So how could Cynthia make Kismet act like a needy infant?**

by people. Another would be a play drive, or a wish to interact with brightly colored toys. The final would be a fatigue drive, or a need to rest if the robot got too much stimulation. When Kismet was switched on, each drive started in a neutral state. The longer these drives went unsatisfied, however, the more intense they became. The more intense they became, the more they motivated the robot to do things that satisfied the drives. If Kismet was switched on in an empty room, for example, it craned its neck left and right, searching for a person or a brightly colored toy in the environment.

~ "Speak to the Robot"

In the spring of 2000, after four years of hard work, Kismet was fully functional. Not only was Kismet a fascinating intellectual and technical exercise, but it actually worked! Cynthia felt proud of her team's accomplishment and of what they had learned along the way. To complete the research for her doctoral thesis, she needed to continue evaluating people's interactions with Kismet. Now that it was completely up and running, she had many research questions to investigate.

Cynthia wanted to find out: How would ordinary people respond to Kismet? Would people consider the robot's behavior life-like? Would the public hate the idea of a robot with "emotions"? Because Kismet did not speak an actual language, what kinds of conversations would people attempt?

When Cynthia shakes a colorful frog toy in front of Kismet's eyes, the action helps satisfy the robot's "play drive."

To find the answers, Cynthia invited many research subjects to meet Kismet. She made sure each person did not have a robotics background. She wanted to make sure the research subjects would offer objective responses to Kismet. She attached a microphone to each person who came into the room, sat him or her down in front of Kismet, and gave a single, simple instruction: "Speak to the robot."

Cynthia videotaped all of Kismet's interactions. When she analyzed them later, she was delighted to find that people had understood Kismet's social cues. Many adults used broad facial expressions and spoke to Kismet in the exaggerated way a mother might speak to an infant. Most subjects knew when it was their turn to speak. One woman asked in a mildly scolding voice, "Kismet, where did you put your body?" In response, Kismet had lowered its head.

A research subject named Rich struck up a lively 25-minute conversation with Kismet. Rich showed the robot his watch. He then explained that it was a gift from his girlfriend and told Kismet he had nearly lost it. Kismet's head leaned toward Rich's wrist to get a closer look, then rose to look into Rich's eyes. Rich left the room feeling that he had forged a real bond with Kismet—even though he knew it was a robot. Cynthia's project was quite a success.

Kismet's behavior surprised many people. One time a reporter came to the AI Lab to interview Cynthia and an MIT postdoctoral student named Anne Foerst. Anne was studying the philosophical and ethical implications of robots. When Cynthia switched on Kismet, nothing happened. Cynthia wondered if she needed to reboot the system. Half-joking, Anne put her face near Kismet's and said, "Why are you doing this? Don't you like me anymore?" As if on cue, Kismet popped into action, looked at Anne, and began babbling sweetly. Everyone laughed. Anne admitted she had not expected Kismet's lifelike behavior to feel so soothing.

~ Pink Ears and Champagne

On May 9, 2000, Cynthia presented her research on Kismet to the MIT professors on her doctoral committee. This was the last grueling exam that she needed to pass to earn her doctorate of science, or Sc.D. degree. Cynthia had written a book-length research paper, or thesis, entitled "Sociable Machines: Expressive Social Exchange Between Humans and Robots." In it, she explained how and why she had designed Kismet. She also presented her analysis of Kismet's videotaped conversations. Now it was time to defend her research in front of a team of professors.

First, Cynthia gave a public presentation of her research. She was surprised to find herself facing a standing-room-only crowd. Clearly, Cynthia—and Kismet—had some fans. Next, she met in

MIT Kismet look-alikes surprise Cynthia after her dissertation defense. Rodney Brooks hams it up on bended knee, while Cynthia's proud mom *(far right)* joins in with her own pink ears.

a smaller room with members of the MIT faculty. This "grilling" session was among the most challenging parts of the exam. The faculty asked Cynthia a series of difficult questions to verify that she really knew her stuff.

Cynthia passed with flying colors. When she emerged from the room, her colleagues were waiting outside to toast her success with champagne. Cynthia laughed at the sight of everyone wearing pink Kismet ears.

Cynthia officially receives her doctoral degree at the MIT hooding ceremony. The ceremony goes back to medieval times, when actual hoods were often used. Hoods today look more like capes.

~ Technology Meets Art

Soon after graduation, MIT Press contracted Cynthia to write a book about her Kismet research. The book, based on her Sc.D. thesis, was published in 2002 and is titled *Designing Sociable Robots.* In the preface, Cynthia reflects on her innovative creation:

> *Kismet connects to people on a physical level, on a social level, and on an emotional level. It is jarring for people to play with Kismet and then see it turned off, suddenly becoming an inanimate object. For this reason, I do not see Kismet as being a purely scientific or engineering endeavor. It is an artistic endeavor as well. It is my masterpiece.*

*"Just my luck.
I finally get
to meet* Tom Cruise

*and he looks
like a* **monster!**"

7

FROM MIT TO THE MOVIES

A t age 32, life was pretty good for Cynthia Breazeal. She was as comfortable studying in an MIT research lab as she was shredding, or snowboarding, on a cold, snowy mountain. The intense pressure of working to earn her doctorate was over. She now had to shift her energies to meet the demands of her postdoctorate year at MIT. Her next plan was to find a job, hopefully as a professor at a college or university. She had applied to a number of schools, including MIT, but was still awaiting an offer.

Then in early 2001, opportunity knocked.

Cynthia got a call from Brad Ball, the president of marketing for Warner Bros., a movie and entertainment company. He had read a recent *Time* magazine article about Cynthia's research with Kismet. He had phoned to find out if Cynthia might be interested in working as a consultant for a movie to be released that summer. Entitled *A.I.: Artificial Intelligence*, the movie was written and directed by Steven Spielberg. The idea for the movie originally belonged to director Stanley Kubrick. But as passionate as Kubrick was about his project, he was too ill to continue developing the story on his own. So he instead became the producer of the film and passed the job of directing to his friend Spielberg, whom he trusted.

In 2001 a glamorous Cynthia *(opposite)* attended the premiere of the movie *A.I.: Artificial Intelligence* in New York City. Above is a detail of Kismet's inner workings.

According to Brad Ball, the movie explored many questions: Is it possible for humans to love robots? What rights and responsibilities should robots have in society? What kinds of friendships might robots of the future develop with each other?

Cynthia was intrigued on many levels. First, there was the thrill of helping out on a Steven Spielberg movie. As a girl, she and her brother had loved many of the classic science fiction movies that Steven had directed, including *E.T.: The Extraterrestrial.* If someone had told the 10-year-old Cynthia that she would one day serve as an expert consultant to promote a Spielberg film, she would not have believed it.

Second, Cynthia was excited about the chance to share her robotics research with a wider audience. Scientists often spend years developing their ideas, only to have them remain hidden in an obscure academic journal. The fact that the movie studio had sought out an MIT consultant meant it was serious about exploring the connections between science fact and science fiction. This would be a fantastic opportunity for Cynthia to educate the public about her mission to build sociable robots.

If someone had told the 10-year-old Cynthia that she would one day serve as an expert consultant to promote a Spielberg film, she would not have believed it.

Third, Warner Bros. would pay Cynthia a consulting fee, which would be a welcome benefit for her.

The prospect was too enticing to pass up. Cynthia eagerly accepted Brad's invitation, then asked him what she had to do. Brad explained that the movie had already been shot and was now being edited. Cynthia's primary responsibility would be to help journalists who would be writing about the film that spring. As reporters researched and wrote articles about the movie, they would have many questions: What exactly is artificial intelligence? What kinds of things can robots do today? How likely is it that robots like David will exist in 50 years? Cynthia was the perfect person to supply the answers.

~ Hollywood Comes to MIT

Soon after, Cynthia flew to Los Angeles to meet with Brad Ball and Kathleen Kennedy, one of *A.I.'s* producers. The three of them hit it off immediately. Cynthia was impressed by Brad and Kathleen's curiosity and enthusiasm. The movie executives, in turn, admired Cynthia's easygoing personality and her passion for her work.

Kathleen, in particular, was well aware of the challenges involved in designing and operating a robotic creature. She had worked with Spielberg on *E.T.* and *Jurassic Park* and had spent much of her career producing movies that brought fictional creatures to life, such as *An American Tail* and *The Indian in the Cupboard.* But the robotic characters in Kathleen's movies had all been mechanical puppets, carefully controlled by special-

In the movie *A.I.*, Haley Joel Osment plays the role of David, a highly advanced robot who longs to be a "real" boy. To create his robotic teddy bear companion special-effects artists created a mechanical puppet.

effects technicians. Cynthia's robots, once switched on, could respond spontaneously to the world, with no additional help from humans. Kathleen was amazed that Cynthia and her team had created Kismet without the benefit of a Hollywood movie budget.

After talking for hours, Kathleen and Cynthia realized that in many ways they had been exploring the same goals for different purposes. Kathleen's movies had created lifelike characters purely for entertainment. Cynthia's lab had done so in a quest for scientific advancement.

Brad told Cynthia that marketing *A.I.* posed a major challenge. Even though the movie had been directed by Steven Spielberg and featured well-known actors such as Haley Joel Osment and Jude Law, Brad and Kathleen were concerned that moviegoers would find the title offputting.

During one of their planning meetings, Brad had a brainstorm: Why not hold the first big press junket at MIT? A junket is a promotional event, sponsored by a Hollywood studio, that gives reporters a chance to interview the cast and creative team of a new movie. A junket hosted by MIT would allow reporters to talk with actors, producers, and real AI scientists as well. Cynthia proposed that her pal Kismet could be made available, too.

Steven Spielberg directs actors Jude Law (right) and Haley Joel Osment on the set of A.I.

On April 30, 2001, hundreds of reporters flocked to MIT's campus. After a tour of the AI Lab, they watched a sneak peek of the film and attended a panel discussion entitled "A.I. the Movie, A.I. the Reality, and A.I. the Future." At this seminar, Kathleen told reporters that any similarities between MIT's research and the movie were purely coincidental. At the time Spielberg had been working on the script and storyboards with Stanley Kubrick, he had no idea that Cynthia was building Kismet.

Actor Haley Joel Osment chimed in on his experience filming the movie. To make his character (David) seem more robotic, he said, he never blinked on camera. Ironically, Cynthia had programmed Kismet with natural blinking behavior to make the robot appear more lifelike.

Grateful to address such a captive audience, Cynthia spoke passionately about why she creates robots. She wants to challenge people's notions of machines, Cynthia explained, while expanding their knowledge of what they could achieve. She asked the reporters, "Must a creature be made of biological materials in order to be real?" Based on her experiences with Kismet, she suggested, the answer might be "No." Although Kismet was clearly a machine, she pointed out, people responded to it with tenderness and compassion.

Cynthia's MIT colleagues spoke, too. Her advisor, Professor Brooks, spoke about how recent medical advances, such as high-tech artificial limbs, were giving humans more mechanical parts—just as robots were being programmed to behave more like humans. Another professor, Sherry Turkle, discussed how children's beliefs have changed over the years about what makes a toy "alive." One hundred years ago, children considered a toy animal "alive" if it simply had wheels. Today, by contrast, a toy animal must be computerized and highly interactive for kids to think of it as "alive."

While the reporters enjoyed meeting real AI scientists, Kismet was clearly the star of the MIT press event. It's one thing to see a make-believe robot on a movie screen—it's quite another to see a real one staring you in the face and making eye contact with you. Kismet's range of facial expressions and babblings charmed the journalists. Many of them said they felt they had met a living creature.

One journalist asked Cynthia, "When do you think artificial intelligence will be a part of our daily lives?"

"In many ways," she replied, "it already is." Every time a person searches the Internet, shops with a credit card, or plays a DVD of a movie, he or she is tapping into technology that grew out of AI research.

~ Directing Steven Spielberg

About a month after the press junket at MIT, Brad Ball called Cynthia to say that Steven Spielberg wanted to meet with her. The director needed to be briefed on the current state of AI so that he could answer journalists' questions on the topic. On the flight to Los Angeles, Cynthia wondered where the meeting would take place: In his office, surrounded by Academy Awards and props from his movies? Or would they discuss robotics over lunch at a trendy Hollywood restaurant?

As it turned out, Spielberg was in the middle of shooting his next picture, *Minority Report,* so their meeting took place at Fox Studios. Cynthia arrived on the set at 9 A.M. A production assistant told her that Spielberg would see her as soon as he finished directing the current scene. On a monitor she saw an actor dressed in black running down a corridor, chasing what looked like a pair of dice. (Later on she would learn that the character in this sci-fi movie was actually chasing his eyeballs!) Cynthia waited patiently as Spielberg filmed the scene over and over again.

When Cynthia finally met the famous director, she marveled at his keen attention to detail. While she waited to answer his questions, she watched him film take after take of a scene, varying the scene slightly each time and observing the impact of the changes.

During their discussion, Spielberg asked Cynthia what kinds of sociable robots might be in our futures. Having been asked these kinds of questions by reporters many times, Cynthia was able to provide clear, concise answers. Spielberg was a good student. Before long he could answer reporters' questions with confidence on his own. Inspired by his conversation with Cynthia, Spielberg later told one reporter that toothbrushes of the future might be able to talk to us. They might, he speculated, "be able to figure out our emotional state and cheer us up if we're feeling blue."

At one point, between takes, the actor from the eyeball-chasing scene came over to say hello. The front of his head was covered with grotesque makeup that made his face look melted. *Who is this guy?* Cynthia wondered. Seconds later, the mystery was solved by these casual words from Spielberg: "Cynthia Breazeal, I'd like you to meet Tom Cruise."

Pointing to his melted face, Cruise said, "As you can see, Steven works me pretty hard."

Cynthia laughed and thought, *Just my luck. I finally get to meet Tom Cruise and he looks like a monster!*

~ A Walk on the Red Carpet

At the end of her tour of duty, Cynthia was invited to the premiere of *A.I.* in New York City. As she strolled down the red carpet in front of the Ziegfield Theater, paparazzi with their cameras and autograph-hungry fans approached Haley Joel Osment, Jude Law, and the other actors. The glamorous scene was something she would never forget.

Back in Cambridge, Cynthia got some great news about her job search: MIT's Media Lab had offered her a job as a professor. She would be staying put at MIT. Cynthia was elated. Jet-setting to California for Warner Bros. had been exhilarating, but she was eager to focus again on robotics research. She had much to do over the next few weeks—coursework to design, students to select for her new group, research papers to write.

There was only one tough part about ending her frequent trips to Los Angeles, and it had nothing to do with robotics.

Becoming a media magnet was exciting, but Cynthia didn't let herself get overwhelmed by the attention.

When *Bobby* and *Cynthia*
returned to the MIT campus,

neither of them imagined
their **friendship** would become
more than that.

PUBLIC AND PRIVATE SUCCESS

I n the early 1990s, Cynthia and a young man named Bobby
Blumofe were both graduate students at MIT. Bobby was working
on his Ph.D. in the Laboratory of Computer Science while
Cynthia was a grad student in the Artificial Intelligence Lab.
Because Bobby and Cynthia's departments were in the same
building, they occasionally ran into each other. They knew each
other's names. Otherwise they were strangers.

Then, during spring break in 1994, Cynthia and a few of her
friends from the AI Lab went on a ski trip to Steamboat Springs,
Colorado. As the boyfriend of one of Cynthia's friends, Bobby
was invited to join them. During the trip, Cynthia and Bobby got
to know each other a little better. They discovered that, among
other things, they shared a passion for skiing and snowboarding.
With the other members of their group, they spent many enjoyable
hours swooshing down the snowy mountain trails. When Bobby
and Cynthia returned to the MIT campus, neither of them imagined
their friendship would become more than that.

Unlike the computer programs that run a robot, however, life is
unpredictable. Fast-forward six years to 2000, when Cynthia was
busy making appearances with her creation, Kismet. One day she
received a phone call from a friend of hers, Steve Keckler, a professor
at the University of Texas (UT) at Austin. Cynthia's success with

In May 2002, Cynthia
and Bobby Blumofe
(opposite) were
married in Beverly
Hills, California. As an
assistant professor,
Cynthia involved her
students in an exciting
project—a terrarium-
based robot *(above)*.

Cynthia takes a break from "shredding" the slopes on her snowboard in Steamboat Springs, Colorado.

Kismet had made her a hot commodity, so Keckler hoped to entice her to interview for a teaching job in the computer science department at UT. As Keckler reeled off the names of former MIT students who had joined the UT faculty, he mentioned Bobby Blumofe.

At the sound of Bobby's name, Cynthia's ears perked up. She asked Steve to "say hi to Bobby for me." Steve explained that Bobby wasn't in Texas at the moment; he was on leave that year to work at Akamai, an Internet company in Cambridge, Massachusetts—right near Cynthia's home.

~ Tracking Each Other Down

Excited about renewing an old friendship, Cynthia e-mailed Bobby. "Hi, Bobby, it's Cynthia," her message began. "Remember me from that ski trip to Steamboat Springs long, long ago? It's great to hear that you're back in Cambridge. Maybe we can get together some time."

In his response, Bobby said it was terrific to hear from Cynthia. But, he explained, he was no longer in Cambridge. Akamai had recently bought another Internet company in San Diego, California. He was now responsible for merging the two firms. He promised to contact Cynthia the next time he was in Boston.

A few months later, Bobby returned to Cambridge for some meetings at Akamai's main office. Out of the blue, he e-mailed Cynthia asking if she was free for lunch. They went to a popular seafood restaurant in Boston, where they talked, laughed, and had a great time catching up. Coming off the success of Kismet— and having just accepted a faculty post at MIT's Media Lab— Cynthia seemed much more self-confident. In fact, Bobby found her "radiant." What a shame they lived on opposite coasts! They wanted to get to know each other better, but it was unlikely they would see each other again soon.

Or so they thought.

~ Together Again

It was just a short time later—the spring of 2001—that Cynthia began flying regularly to California for her consulting work with Warner Bros. The frequent business trips to Los Angeles gave her a chance to visit Bobby in San Diego, just two hours away by car. Little by little, a relationship blossomed.

At the end of May, Bobby invited Cynthia to San Diego to surf. Though a lifelong jock, Cynthia had never learned to surf. During her college days in Santa Barbara, she had avoided battling the mobs of talented surfers that made it riskier for newbies to learn.

Now, Cynthia had a wonderful time with Bobby in the waves off the San Diego coast. They enjoyed each other's company and eventually became romantically involved. In July, however, Cynthia was scheduled to begin teaching at the MIT Media Lab. For their relationship to survive, one of the two would have to move to the other coast.

Then tragedy brought Cynthia and Bobby together. On September 11, 2001, Daniel Lewin—Akamai's cofounder and inspirational leader—was aboard one of the planes that terrorists crashed into the World Trade Center in New York City. Lewin's death was a devastating blow to morale at Akamai. It also couldn't have come at a worse time for the company's fortunes. Internet companies that had thrived just a few years earlier were now struggling to survive in the bad economy. To cut back on costs, Akamai closed its San Diego office and transferred Bobby back to Cambridge. Although they deplored the circumstances that had reunited them, Cynthia and Bobby were happy to be living in the same city.

"Hi, Bobby, it's Cynthia," her message began. "Remember me from that ski trip to Steamboat Springs long, long ago?"

~ Back to Work

As Cynthia's personal life blossomed, so did her professional life. In her new role as an assistant professor, Cynthia formed the Robotic Life Group at MIT's Media Lab. Its mission is to "build cooperative robots that could work and learn in partnership with people." By creating socially intelligent robots such as Kismet, Cynthia hoped to change the view that robots are nothing more than high-tech tools. For Cynthia, creating robots went beyond an intellectual exercise: One day, she dreamed, the robots that she and her students created would become part of daily life for millions of people.

Although Cynthia had served as a teaching assistant before, this was the first time she had been given total control of teaching a class. How could she inspire her students to do great things?

At the Robotic Life Group's research lab *(right),* inspiration for robot design can come from anywhere—plants, primitive animals, or sophisticated animals or people. According to Cynthia, "My lab is in a constant state of creative chaos as we design our next inter-active robot."

The first time Cynthia's students met for class, she broke the ice by asking everyone to name their favorite robots from movies or TV. Cynthia spoke of her passion for R2-D2 and C-3PO from *Star Wars*. She also mentioned her fondness for Data, the android from *Star Trek: The Next Generation*. Her students named the science fiction films that had spurred them to pursue careers in robotics: *Silent Running, Blade Runner, Bicentennial Man*. Cynthia was encouraged to discover such an eager group of budding roboticists.

Given almost unlimited freedom by the Media Lab to research and teach a subject of interest to her, Cynthia challenged her group to come up with a bold and innovative first project. Her goal was to build a cooperative spirit by motivating her students to pool their talents. Instead of creating an elaborate robot that would be seen by only a handful of MIT scientists, Cynthia wanted her research group to create something extraordinary for a huge audience.

What lay behind her thinking? In Cynthia's experience, bright students rise to the occasion when faced with a stimulating challenge. Knowing that their work would be seen by thousands of people— and that it might be reviewed by the national media—would motivate the students to strive for excellence. A large project would also allow Cynthia's students to collaborate with other professors and students at the Media Lab.

~ The SIGGRAPH Challenge

For all these reasons, Cynthia applied to SIGGRAPH (Special Interest Group for GRAPHics) 2002, one of the top computer graphics conferences in the world. Tens of thousands of artists, programmers, producers, and professors attended the conference each year. Cynthia was confident that many technology experts would see her students' creation.

Cynthia scrutinized the SIGGRAPH submission categories. Two of them—the Art Gallery and the Computer Animation Festival—focused exclusively on computer art and animation. Those would not work because Cynthia's Robotic Life Group would be creating something three dimensional.

The conference's Emerging Technologies showcase, on the other hand, seemed like a perfect fit. The directors of this category were looking for examples of research that explores the relationship between humans and machines. It wouldn't be easy, but Cynthia thought her students could create something convincing for this category. As Robotic Life Group member Dan Stiehl later commented, "There's nothing like a big demo and a lot of stress to pull people together."

Once Cynthia learned that her group's application to SIGGRAPH 2002 had been accepted, her team had only about one year to complete its project. That was precious little time for all they had to do. First, they had to figure out a great concept that would show off their creativity and technological mastery. Then they had to construct and program everything. Finally, they had to ship the entire creation to the SIGGRAPH conference site in San Antonio, Texas.

The team excitedly brainstormed ideas. What type of robotic creature would be fun and unusual, alien yet lifelike? What kind of environment might this mechanical being inhabit? What sort of materials would make the robot look and feel alive? As the group explored answers to such questions, Cynthia laid down her one rule about designing robots: No robot should look or behave exactly like an existing animal or plant. "Keep things fun and fanciful," she urged her students. That directive arose from Cynthia's deeply held belief that robots should be designed as if they are a unique species.

During an early idea session, one student suggested that the group create a tank filled with mineral oil and "glowing goo." Inside would be a lifelike robot that resembled a sea anemone. In the real world, a sea anemone (a-NE-mon-ee) is a colorful, ocean-dwelling animal that looks like a harmless ocean flower. It attaches itself to coral, rocks, or the seafloor, where it uses poisonous stingers on its tentacles to catch small fish, worms, and crustaceans.

The robotic anemone might resemble its biological counterpart, but it would be programmed to interact with people. When

> The team excitedly brainstormed ideas. What type of robotic creature would be fun and unusual, alien yet lifelike?

SIGGRAPH attendees walked around the outside of the tank, the robot would follow their movements. It would even try to touch them through the glass!

Everyone loved the idea of a robotic ocean creature. The concept of the fluid-filled tank, however, presented practical problems. Even with careful planning, Cynthia's students knew, experimental robots often break down. An anemone-like robot submerged in a vat of mineral oil would be almost impossible to repair.

The Media Lab makes its home in the Weisner Building, shown here at night. The building was designed in 1985 by the world-famous architect I. M. Pei [PAY].

Still enamored of their anemone, the group tried to think up another approach. What if they built a terrarium instead of an aquarium? A terrarium is an enclosed glass container that a biologist might use to keep lizards, toads, or turtles. A terrarium with open sides would allow the public to interact with the robot and its environment. To model an aquatic environment, the group could add a waterfall and a pond design.

As the group's goals became more ambitious, the artificial environment it was envisioning grew larger in size. Once merely the size of a small coffee table, the habitat they now foresaw measured 7 feet wide × 7 feet long × 10 feet high.

Cynthia loved the interactive terrarium concept, particularly because it involved an anemone-like creature. Most people tend to visualize robots as mechanical people with two arms and two legs. Creating robotic tentacles was a very interesting thought.

Each of Cynthia's students was also required to conduct an individual research project. She was pleased to see that these efforts generated ideas that added depth to the overall project. The group worked hard to accommodate each new notion. Doctoral student Josh Strickon, for example, wanted to develop an "interactive show-control system." Josh would program the system to automatically coordinate such entertainment elements as lighting, sound effects, and music. To address Josh's request to make lighting changes a key element in the project, the team decided to give the terrarium two cycles—night and day.

Adding a night phase had two other advantages.

First, more nocturnal robotic creatures could be incorporated. Glowing tubeworms, for example, could appear as the anemone "slept."

Second, if the anemone broke down (okay, *when* the anemone broke down) during the conference, the team could use the night phases to discreetly make repairs.

For many weeks the anemone-like robot was known simply as "the water creature." Then, student John McBean suggested a new name: Public Anemone. Everyone loved it. The name combined the robot's biological inspiration with the fact that their creation would be seen by a large number of people. What's more, it spoofed the name of a popular rap group, Public Enemy.

~ Blood, Sweat, and Gears

Building the robot and its terrarium environment took Cynthia's group about eight months. At any given moment, each team member—there were now more than 10 of them—worked on a different part of the project. While one student figured out the robot's movements, another molded its elastic skin. As some tackled the problem of designing a homemade waterfall, others toiled in front of computer screens, writing software to control the robot's movements.

The team members developed their initial model based on sketches and 3-D computer models. Because they were creating something no one had built before, it wasn't easy to track down the right supplies. Special-effects supply houses were often a big help.

Designing the anemone's metal arm and tentacles also posed a wealth of challenges. How could the team make the robot's movements fluid and natural? Team members attached servo

motors between each of the robot's sections. A servo motor is a device that uses electricity rather than mechanical force to change the position of a motor. The servo motors served nicely to smooth out the robot's movements.

Daunting too was the artificial skin design. Student Dan Stiehl had studied movie-makeup techniques since high school, so he guided others through the process. First, he created a clay model of the skin. Then he made a mold of the model and filled it with silicone rubber. Once this material was dry, Stiehl painted it with bright lifelike colors.

When it came time to construct the terrarium, the team relocated to a nearby warehouse to find more room. Here, Public Anemone really began to take shape. To create the illusion of a rocky terrain, the crew built a frame made of metal and plywood, wrapped it in chicken wire, and covered it with polyurethane foam. When the foam hardened, they carved it to look like rocks.

Getting the waterfall to flow naturally into the pond took a lot of trial and error. At first the water gushed almost straight out, forcing the team to construct a new edge to make the water flow

Public Anemone's hand-painted synthetic skin *(top)* is made of highly elastic silicone rubber. It's designed to fit over the robotic creature's mechanics *(above)* in a way that results in lifelike, natural movement.

downward. Finally, the team added some dried and some painted flowers plus patches of actual moss to blur the line between fantasy and reality.

On the surface, Public Anemone looked nothing like Kismet. But both were designed to be autonomous robots, so their programming shared some key features. Both robots' programming contained artificial "drives," or desires. Kismet's software, for example, made the robot seek out people when its social drive was low. Similarly, Public Anemone was programmed with drives to support the illusion that it was a lifelike creature. It had strong desires to complete its chores, including watering nearby plants and bathing itself in the waterfall. If the robot succeeded in these basic tasks, its software instructed it to interact with nearby humans. If not, the robot was programmed to exhibit a self-protective response.

~ Sights, Sounds . . . and Smells!

As spring turned into summer, it was time to finish the project and prepare it for presentation at the SIGGRAPH conference. Cynthia's team had constructed an elaborate and delicate physical world, so transporting their materials from Cambridge to San Antonio would be an ordeal. Two weeks before the conference, everything was carefully packed into three huge crates and loaded onto two trucks. "In the warehouse," Dan recalled, "we couldn't fit the largest crate into the freight elevator, so we had to take the elevator doors off." The team didn't want to leave anything to the last minute, so they even packed up rolls of real moss.

Once the crates arrived in Texas, setting up the terrarium became another test of wits. At times it felt like assembling a 1,000-piece jigsaw puzzle. In addition to erecting the rocklike terrain, the team had to set up the visual sensors—hidden video cameras that tracked people's movements as they interacted with the robot.

Once the waterfall and pond were in place, the students made a delightful discovery: When the moss became wet, the terrarium smelled like a swamp. Now their make-believe world appealed to yet another sense! The only downside was that the mossy water had to be changed nightly, otherwise their charming world would grow into a disgusting, grungy mess.

At the conference, Cynthia encouraged her students to experiment with Public Anemone and make last-minute improvements. Bill Tomlinson, a Ph.D. student from the Media Lab, offered a fun idea for enhancing the robot's interactivity: If someone got too

Members of Cynthia's Robotic Life Group pose in front of Public Anemone and its environment at a technology conference in San Antonio, Texas.

close to the anemone or made a sudden movement in its presence, the robot should make a rattlesnake sound and abruptly withdraw its body. The team found this proposal intriguing. A few hours of clever programming later, Bill's ideas became part of the demo.

~ A Big Risk Pays Off

The Public Anemone project became one of the most popular exhibits at SIGGRAPH 2002. The conference chairman praised the robot for appealing to a wide audience; both his mother and his 6-year-old son, he reported, were dazzled by it!

During the night phase of the robotic terrarium, visitors would tap on drum crystals to create a rhythmic percussion sound. The drumbeat set off a light display that glowed against the terrarium wall.

Because the experience was different each time, many people returned to visit the terrarium again and again. Cynthia was especially proud of her students' ability to remain creative and cooperative as they juggled hundreds of details. When other professors congratulated Cynthia on the multimedia project, she quickly diverted the praise to her students. Without their ingenuity and dedication, Public Anemone and its world would never have come to life.

~ Another Happy Ending

Cynthia and Bobby at the White House in 2003, where she was honored as a finalist in the National Design Awards.

In 2002 another joint venture brought Cynthia great joy. Bobby Blumofe and Cynthia returned to one of their favorite cities, San Diego, to celebrate their one-year anniversary as a couple. They surfed, strolled along the beach, and dined at their favorite restaurants.

Then Bobby surprised Cynthia: He presented her with an engagement ring and a marriage proposal. Knowing their relationship was strong, she confidently said "Yes."

On the afternoon of June 7, 2003, Cynthia and Bobby were married in Beverly Hills, California. After the wedding, the newlyweds honeymooned in the Polynesian islands of Tahiti, Bora Bora, and Huahine. The islands were the perfect place to relax, scuba dive, and surf. But the best was yet to come. On March 12, 2004, Cynthia Breazeal produced her great-

Cynthia took 4-month-old son, Ryan, with her to the city of San Gimignano in Tuscany, a region of Italy. She lectured a summer school course there on the subject of human and robot interaction.

est creation ever. That's the date when proud parents Cynthia and Bobby welcomed their baby son, Ryan, into the world.

During her pregnancy, Cynthia had prepared for her new role by reading a number of books. She thought she understood the basics of soothing a crying newborn, burping it, and so on. Then Ryan arrived and—surprise! Cynthia and Bobby realized they still had much to learn. So they hired a "night nurse" for a few weeks to provide support and advice.

In the months that followed, Cynthia found that juggling the demands of her job at MIT with her responsibilities as a new mother was a major challenge. On especially stressful days, her son's laughter helped put things in perspective. The first time Ryan laughed, Cynthia recalled, was "simply magical. He was so purely happy."

Even though
Cynthia came from the
world of science *and*
Stan Winston came from the world

of **entertainment**,
she wondered if they could
pool their talents.

Building a Bridge between Worlds

9

Cynthia and special-effects creator Stan Winston *(opposite)* stand by their "strange and wonderful off-spring." A close-up of Leonardo's eye *(above)*.

Professors and students rarely have the luxury of working on a single project at a time. In the fall of 2001, as Cynthia's students were developing the Public Anemone demo for SIGGRAPH, they also began work on another sociable robot.

Cynthia dreamed of creating a robot that would take Kismet's technology to the next level. Kismet's natural motion and real-time interactivity had broken new ground. For this next machine to look and act more like a living creature, it would need more realistic facial expressions. It would also require more sophisticated sensors, motors, and software. Finally, it would need to interact with people as individuals by recognizing faces and remembering specific interactions.

In short, to be more socially intelligent, the robot would need to be more lifelike.

Cynthia was concerned that her lofty goals clashed with her lean budget. Kismet's materials had cost about $25,000. And that didn't take into account the expense of people's time. Labor is almost always the most expensive part of any robot-building venture. Including labor, Kismet cost more than $100,000. Constructing the type of realistic robot she was thinking of could take more than a million dollars. Where would this money come from?

Media Lab research is funded mostly by government grants and corporate sponsors. Getting funded, however, can be challenging because of the stiff competition coming from researchers everywhere. Not all grant requests, or proposals, are accepted. But even with unlimited funding, Cynthia's team lacked the technical know-how to create realistic-looking robots. Kismet was cute and charming, but it clearly had more brains than beauty. Where in the world could she find special-effects experts to help her?

Perhaps the answer was back in Hollywood. In 2001, while watching the movie *A.I.*, Cynthia was impressed by the movie's robotic characters. Some weren't machines at all—they were actors wearing makeup. Other characters existed only as onscreen computer-generated images. But a few, such as Teddy, the hero's robotic teddy bear, were mechanical wonders. Teddy's facial expressions were remarkably lifelike. His motions were guided by puppeteers who used high-tech remote-control devices. When one of Teddy's puppeteers lifted his own right arm, for example, Teddy the puppet performed the same motion.

Stan Winston Studio, the company that created *A.I.*'s robotic characters, had succeeded in making Teddy appear to be a living, breathing creature. Even though Cynthia came from the world of science and Stan Winston came from the world of entertainment, she wondered if they could pool their talents.

Kismet was cute and charming, but it clearly had more brains than beauty. Where in the world could she find special-effects experts to help her?

Cynthia's involvement with *A.I.* began after the film was shot, so she had never met Stan Winston and his team. She knew Stan only by his reputation. He was one of the top special-effects experts in Hollywood, and had won Academy Awards for his work on *Aliens, Terminator,* and *Jurassic Park.* Cynthia thought it would be rude to call him out of the blue. Instead, she asked *A.I.* producer Kathleen Kennedy to arrange a meeting with Stan on her behalf. Kathleen gladly agreed.

~ Stan's Studio

Stan Winston Studio (SWS) occupies a 35,000-square-foot building in the San Fernando Valley. Stan was out of town the first time Cynthia visited SWS in the summer of 2001. But he had arranged for her to meet with Lindsay MacGowen, the special-effects artist who designed Teddy's appearance.

Lindsay gave Cynthia a tour of the studio, showing her the various versions of Teddy that had been used for *A.I.* One Teddy had been featured in close-ups with human actors; another was a "stunt Teddy"; still others were only half-Teddies. The Teddy carried by Haley Joel Osment's character weighed more than 30 pounds. It could be made to curl up, wiggle its nose and ears, and grab objects. This robot had 50 servo motors in its body, Lindsay explained, almost half of them in its head. *No wonder Teddy could make such great facial expressions*, Cynthia thought. She was awed by the SWS creations.

Lindsay was equally impressed by Cynthia's talents. Cynthia and Lindsay discovered they shared an interest in science fiction books and movies. Cynthia mentioned her interest in building another robot that would be as sociable as Kismet but more life-like. Would SWS like to collaborate on it? Intrigued, Lindsay suggested they meet with Stan to discuss the possibility.

A few weeks later, Cynthia returned to SWS. She asked Stan Winston, "How would you like to build a real Teddy?" If SWS would design—and pay for—a robotic character, Cynthia volunteered, MIT would supply the sensors and software that would enable the robot to see, hear, speak, and even touch. In exchange for this SWS financing, MIT would share the secrets of its state-of-the-art AI technology. Excited by the offer and impressed with Cynthia's enthusiasm for her work, Stan immediately accepted.

Several weeks later, Cynthia returned to SWS with several of her students. As SWS technician Richard Landon recalled, "For two days, we shared knowledge. The idea was, 'You tell us what you do; we'll tell you what we do, and we'll find a common mid-ground.'"

The two teams toyed with the idea of creating some kind of "performance art"—a type of show that was light on plot and heavy on high-tech dazzle. While this possibility seemed interesting, they continued to brainstorm. Next, they briefly considered turning one of the animatronic Teddies from *A.I.* into a Kismet-like autonomous robot. This idea was discarded for technical reasons: The type of electrical system used for Teddy, Cynthia learned, differed completely from the type used by MIT's AI experts. This meant that Teddy's gears, motors, and wires were nothing like those used for Kismet.

Then someone suggested that the two teams create a new robotic creature from scratch. Everyone was excited about this idea. As they spoke about the possibility, the steps of their collaboration began to gel. The SWS team would start by designing a character the way they usually did. First, they would make sketches of a possible character. Once Stan approved the sketches, the team would make a 3-D sculpture based on them.

After the sculpture got Stan's okay, an animatronic version of the character would be built. As usual, one of Stan's puppeteers would use a remote-control device to dictate the motions of this electronic puppet. When the SWS team got to the animatronic puppet stage, however, it would include *both* types of electrical systems: theirs and MIT's. This would allow the SWS team to use its electrical system to test and demonstrate the robot's abilities. Cynthia's team could then remove this gadgetry to embed, or install, sensors and program the robot using MIT's technology.

By the end of the two days, both teams were thrilled about the collaboration. Richard Landon's other experiences with scientists had been nowhere near this productive. Those other scientists seemed more concerned about protecting their research and discoveries than on benefiting from collaboration. Cynthia, by contrast, was honest, open, and eager to listen to other people's ideas. "When you're not trying to trick people out of information," Richard said, "it opens the whole thing up. Then people want to play and be creative."

~ Cynthia's Advice

Before Stan's team designed the new robot, Cynthia offered a few guidelines based on her experience.

First, don't make the robot's face look human. In the 1980s, Japanese roboticist Masahiro Mori had conducted research showing that the more humanlike features a robot has, the more appealing it is to people. But Masahiro found that there is a limit to this appeal— a concept he called the "uncanny valley." (*See box, page 49.*) Fortunately, this request wasn't a problem for Stan's team, which was fond of creating fantastic fantasy characters.

Cynthia's second guideline suggested that the robot not resemble a real animal. Dogs and cats already exist, so what's the point of making robots that look and act just like them? Borrowing individual features from different animals, however, was fine. "Give the character its own reality," she advised.

Third, Cynthia requested that the character be childlike. Because AI technology has not yet progressed to the point where scientists can craft believable adult robots, creating a robotic "child" seemed the smart way to go. In other words, it would be important for the character's brain and body to be in sync, or on an equal level.

Many think Leonardo is the most expressive robot that exists today. Of its 61 degrees of freedom, 32 of them are in its face alone.

~ Lovable Leonardo

In May 2002 the SWS team came to MIT to unveil the character that they had created. Cynthia was instantly enchanted. The new creature was cute and cuddly. It was furry, stood two and a half feet tall, and stared out at the world through big brown eyes. With its fuzzy ears nearly as long as its arms, it resembled a title character from Steven Spielberg's 1984 movie *Gremlins*.

Leonardo (minus its fur coat), with name-sake Leonardo da Vinci *(below)*. Centuries ago, the brilliant artist-genius sketched what many believe is the first known design for a humanoid automaton, or self-moving machine.

Stan Winston explained that they had named the new creature Leonardo in honor of Leonardo da Vinci, the famous Renaissance scientist, artist, and inventor. *What a perfect name,* thought Cynthia. *I wonder if they know that da Vinci is one of my heroes?*

Leonardo had a large head and belly—mainly because Cynthia's team needed to hide cameras, microphones, and other equipment inside the robot's body. It also had a thick, soft fur coat that could be removed, so Cynthia's team could work on the machinery beneath it. The reason the fur looked so lifelike was that it came from goats and yaks. It had been stitched into the robot's coat one strand at a time.

When an SWS puppeteer made Leonardo move, the illusion was remarkable. With more than 60 degrees of freedom, or directions in which its parts could move, the robot seemed exceptionally lifelike. Leonardo, soon nicknamed Leo, had more intricate machinery than the animatronic dinosaurs that SWS had created for the *Jurassic Park* movies. More than half of Leo's motors were in its head, enabling its eyes to roll in all directions. Its ears could even twirl in opposite directions. The robot could grin, grimace, scowl, and smirk.

But Leo was no perfect puppet. It was a remarkably sophisticated machine, but in the end it was just that, a machine. It occasionally did what machines can do: malfunction and break down. When mechanical mishaps occurred during Leo's construction phase, the SWS team dubbed them the "wagga waggas."

By summer 2002 it was time for SWS to hand over Leo to the MIT Media Lab. Cynthia and her group gradually transformed Leo from an animatronic puppet into an autonomous robot. They inserted various sensors into Leo's body so that it could see, hear, and touch the world around it. They also wrote software programs that would allow Leo to respond to stimuli with appropriate movements, gestures, and expressions. The programming for Leo's "eyes" enabled it to scan a room in search of people. From the robot's point of view, it was looking around for anything that moved and had two eyes and a face.

In addition to giving Leo the ability to see, hear, and touch, Cynthia's team made certain practical adjustments. Mechanical humming noises in the SWS robot version were okay for a movie set because they could be edited out of the movie's soundtrack. But if Cynthia's version of Leo buzzed or hummed during face-to-face interactions, the noise would remind people that Leo was a machine. To keep the gear and motor sounds as low as possible, Cynthia had asked Stan's team to use high-end motors during the construction phase.

> Cynthia and her group gradually transformed Leo from an animatronic puppet into an autonomous robot.

Other programming that Cynthia's team began at this time included making Leo able to recall specific interactions. The idea was that if someone were to behave kindly toward Leo, its computerized memory would record the exchange. If the person visited again, Leo would likely smile and appear friendly. If, on the other hand, someone were to threaten Leo on an initial visit, the robot would act defensively the next time that person approached.

Cynthia's team also concentrated on giving Leo a more sophisticated sense of touch. Eventually, the robot will be able to detect the different shapes and densities of items it holds in its hands.

Leonardo keeps its sights on a yellow ball in his left hand. The Media Lab is working on a sensory touch system under the robot's synthetic skin that allows Leonardo to respond to human touch.

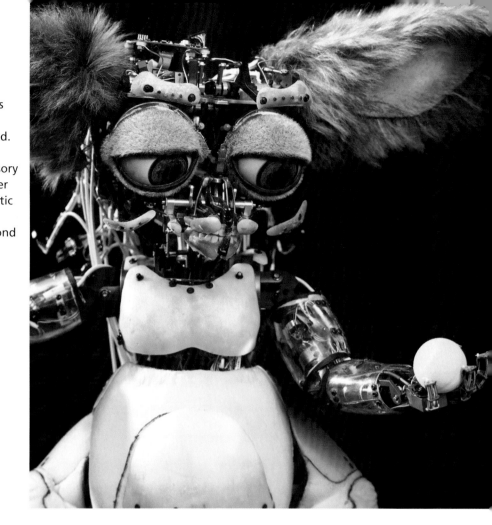

~ Leonardo the Living

The first time the SWS team saw Leonardo in action at MIT, they couldn't believe their eyes—or the robot's eyes, to be precise. Even though Leonardo's movements were still a work in progress, Stan Winston, Richard Landon, and Lindsay MacGowen were amazed by the creature's ability to make and maintain eye contact.

In October 2002, *Scientific American Frontiers*, a TV series hosted by actor Alan Alda, introduced Leonardo to the public. Impressed with Leonardo's capabilities, he congratulated Cynthia and Stan on their creative collaboration. Their give-and-take kept the discussion lively. After describing Leonardo's long-term goals, Stan joked, "And I'm going to take credit for it all." Everyone laughed as Cynthia replied, "Oh, no you don't!"

~ Just a Trick?

Not everyone was equally mesmerized. Computer science pioneer Marvin Minsky, a cofounder of MIT's first AI Lab in 1959, commented: "My objection to Leonardo is [that] it's just a trick. It doesn't really have emotions. It just knows how to fool you into thinking it does. Cynthia's an excellent engineer, but her work doesn't explain how emotions work."

Marvin's criticism did not surprise Cynthia. After all, he had a history of upsetting AI scientists: "AI [research] has been brain-dead since the 1970s," he once remarked. Cynthia also knew that good science always includes questioning and debate.

However, Cynthia felt that Marvin had missed the point of her research. In her view, Leonardo's social skills and "emotions" help advance the human-robot relationship. Cynthia's goal was— and still is—to create robots that can cooperate with people as partners. By providing robots with emotional intelligence, Cynthia feels that these machines will communicate more effectively and personably with others.

~ Leonardo's Influence

Cynthia's team continues to improve and expand Leonardo's abilities. As the robot's capacity to communicate with others becomes more sophisticated, it will continue to inspire collaboration among scientists at the Media Lab. To recoup some of the steep costs involved, Stan Winston is exploring the possibility of using Leonardo in a movie.

This much is certain: Cynthia and Stan take great pride in their work. As Stan once enthused, "Leonardo will affect many people's lives by virtue of the fact that he exists. He's not science fiction but science reality. He's a strange and wonderful offspring."

Cynthia **smiled**
when she saw the photos
of Kippit,

the simple robot
that the **girls**
had designed.

THE ROBOTS ARE COMING!

10

I n April 2004, while Cynthia was on maternity leave from MIT, she received a surprising e-mail from a man named Vernon Ellinger. His 10-year-old daughter, Katie, and her friend Claire Velez were big fans of Cynthia's work in robotics, Vernon wrote. *How in the world did these kids hear about me?* Cynthia wondered. The e-mail explained that Katie and Claire had done a report on Kismet for their elementary school science fair. Not only that, but they had created their own emotionally expressive robot, named Kippit.

The girls lived nearby in Massachusetts and were on a break from school, so Vernon asked Cynthia if she would be willing to meet Katie and Claire somewhere in Cambridge. *How cool!* Cynthia thought. She e-mailed back an enthusiastic "Yes!"

A few days later, Cynthia—with baby Ryan in tow—met Katie and Claire in a cafe in Harvard Square. The girls proudly showed Cynthia the posters they had made for their science project. Cynthia smiled when she saw the photos of Kippit, the simple robot that the girls had designed. It had three different facial expressions—happy, mad, and neutral. One poster explained that Kippit's mouth and eyebrow movements were controlled by electromagnets. The girls' poster said, "An electromagnet is a wire with electricity running through it, coiled around iron, in our case

Budding scientists Katie Ellinger *(opposite far left)* and Claire Velez *(sitting next to Katie)* meet their hero Cynthia and tell her about Kippit, the expressive robot they created. Above, Cynthia signs one of the girls' posters.

In this rear view of Kippit (*right*), you can see the wires that use electro-magnetic power to control the robot's facial expressions. In the front view (*below*), one of the girls is preparing to make the robot smile by turning on a switch, which was actually Kippit's nose.

a nail." By touching each electromagnet to a power source (four D batteries), the girls were able to change the positions of Kippit's mouth and eyebrows.

Impressed, Cynthia asked the girls how they had made the electromagnets. The more times a wire is wound around a piece of iron, the girls had learned in science class, the stronger the electromagnet will be. Katie and Claire explained that Katie's dad had helped them by inserting a nail into the end of an electric drill. The girls then attached a copper wire to the nail and turned the drill on. As the nail spun around, the copper wire wrapped around it hundreds of times.

Katie and Claire had all sorts of questions for Cynthia. Most important, they wanted to know if she had been interested in building robots when she was their age. Cynthia grinned and told them that at age 10 her passion had been soccer. The girls were surprised to learn that Cynthia did not build her first robot until she was in graduate school. They loved hearing that Kismet was inspired by C-3PO and R2-D2 from *Star Wars*. Many scientists, Cynthia pointed out, are similarly inspired by science fiction.

On the trip home, Cynthia wondered what careers Katie and Claire might pursue when they grew up. Would they become roboticists? Cynthia had enjoyed science when she was 10, but she never dreamed of becoming a "creature creator." By the time these girls were in college, the study of robotics would probably be vastly different.

It was a little unnerving for Cynthia to think of Kismet as "old-fashioned."

~ Robots of Tomorrow

Cynthia thinks a lot about the future. How long will it be before every family has at least one robot in their home? What will these robots look like? What kinds of activities will robots and humans perform together? How might new developments in technology revolutionize the world of robotics? Will laws have to be written to protect the rights of robotic creatures?

But Cynthia doesn't just sit back and ponder the future. She follows the advice of technology visionary Alan Kay, who said, "The best way to predict the future is to invent it." At MIT's Media Lab, Cynthia and her team are busy inventing prototypes for robots of the future. Given MIT's impact on technology, she knows that her work has the potential to influence how robots will be integrated into our lives. Cynthia believes that robots have the potential to enrich important aspects of our lives, such as our health and well-being, much the way pets do. Just because the word "robot" comes from the Czech word for "slave labor" does not mean that robotics must move in a utilitarian, or strictly practical, direction.

Dear Dr. Breazeal,
Thank you so much for meeting with us in Harvard Square. We appreciate your views and opinions about robots and other types of science. We learned that theory and hands-on science create a cycle because when you work on one, it helps to improve the other.
You taught us a lot about Kismet and your other robots. We hope you learned a lot about Kippit.

Your Fans Forever,
Katie Ellinger
Claire Daley

After meeting Cynthia, Katie and Claire wrote her this letter of thanks. Sharing insights and opinions with Cynthia helped the girls understand more about robotic science and the life of a scientist in general.

~ Robotic Companions

Cynthia is not the only one who believes that humans and robots can enjoy each other's company. According to a 2003 survey conducted by the United Nations, more than a third of the robots in the world are designed for entertainment purposes. The AIBO, for example, is a robotic dog produced by the Sony Corporation.

It comes with software that enables it to learn tricks and develop its own personality based on the interactions it has with its owner. Unfortunately, each AIBO costs nearly $2,000.

To manage the costs of research, roboticists around the world are exploring practical uses for robots in everyday life. Health care could be the first area to see the expansion of robot use. In Japan, where the elderly population is outgrowing the number of available caregivers, Cynthia wonders if sociable robots might come to their aid. A socially intelligent robot could be taught to cook and clean and could provide companionship. It could also learn an individual's preferences. The effect would be to improve people's quality of life.

Aware that most people are wary of giving control of their lives to a machine, Cynthia suggests a cooperative relationship with robots. Rather than thinking of robots as tools, humans could collaborate with them as companions to tend a flower garden or cook a meal for friends.

~ A New Way to Preserve Memories

Cynthia Breazeal is constantly on the lookout for innovative ways to apply robotics technology. Sometimes these ideas come from unexpected places. One of the companies that funds research at the Media Lab is a large greeting cards company. Why would it care about robots? The company is interested in connecting people across distance and time. The company is also seeking new ways to help people make and preserve memories.

Sociable robots, Cynthia believes, may eventually help support this mission. "What if a robot lived with a person from childhood through adulthood?" she asks. Through interactions with that person, the robot would develop a real sense of his or her personality and experiences. Then at the end of the person's life, the robot could be passed on to children and grandchildren. In theory, the person's great-great-great-grandchildren could ask the robot what their relative was really like—what the person did for fun, what the person was afraid of, and so on.

Although the NS-5 robot from the movie *I, Robot* will not become a reality anytime soon, it is an example of science fiction that can inspire one's imagination. The movie is loosely based on a story by author Isaac Asimov, who wrote many stories about robots. He imagined a world where robots looked after the well-being of humans.

~ A Model for the Future

It's exciting to imagine what future generations of scientists will bring to the field of robotics. Cynthia Breazeal is proud to be among those leading the way. From childhood, her curiosity and perseverance have allowed her to break through what others saw as obstacles. Confident that everything she learns has the potential to produce new ideas, Cynthia has always embraced new experiences.

Challenge and change do not daunt her; instead, they are calls to action. And machines such as Cog, Kismet, and Leonardo are more than mere robots; they are proof that successful scientific pioneers—like Cynthia Breazeal—truly exist.

TIMELINE OF CYNTHIA BREAZEAL'S LIFE

1967   Cynthia is born on November 15 in Albuquerque, New Mexico.

1970   The Breazeal family moves from Albuquerque to Livermore, California, after Cynthia's father is transferred there by his company.

1977   Cynthia sees the original *Star Wars* movie with her family and becomes enchanted with its robot heroes R2-D2 and C-3PO.

1985   Cynthia graduates near the top of her class at Granada High School in Livermore. After briefly considering a career in professional tennis, she decides to go to college instead. She enters the undergraduate engineering program at the University of California, Santa Barbara (UCSB).

1988   During the summer, Cynthia interns for the Xerox Corporation at its office in El Segundo, California. There she uses her engineering skills to test microchips.

1989   Cynthia graduates with high honors from UCSB with a bachelor's degree in electrical and computer engineering. She decides to pursue a career in robotics.

1990   Cynthia moves to Cambridge, Massachusetts, to attend graduate school at the Massachusetts Institute of Technology (MIT). She studies robotics with Professor Rodney Brooks at MIT's Artificial Intelligence Laboratory. At age 22, she builds and programs her first robots.

1993   Based on her research with the insect-like robots Hannibal and Attila, Cynthia is awarded a master's degree in electrical engineering and computer science from MIT. Professor Brooks's students begin building Cog, a robot with human-like capabilities. Cynthia becomes the project's chief architect and programmer.

1997   Cynthia begins work on Kismet, an emotionally intelligent robot.

2000   Cynthia defends her doctoral dissertation (or thesis) on Kismet and
        earns a doctor of science degree from MIT. Her research attracts the
        attention of the media and *Time* magazine publishes an article about her.

2001   As part of the marketing campaign for the movie *A.I.: Artificial Intelligence*,
        Warner Bros. hires Cynthia as a consultant to help educate the public
        about robotics. MIT hires Cynthia as an assistant professor in its
        Media Laboratory, where she forms and directs the Robotic Life
        Group. Cynthia and her students collaborate with Stan Winston
        Studio, a special effects company in Hollywood. Together, they build
        Leonardo, a socially intelligent robot.

2002   Cynthia's first book, *Designing Sociable Robots*, is published.

2003   Cynthia marries Bobby Blumofe, a computer scientist she met when
        they were graduate students at MIT. Kismet is officially retired and
        goes on display at the MIT Museum.

2004   Cynthia's son, Ryan Fulton Blumofe, is born on March 12 in Cambridge.

2005   Cynthia continues her work as director of the Robotic Life Group at
        MIT's Media Lab.

About the Author

Through some ghastly but marvelous error, Jordan D. Brown was hired to write *Robo World*, even though he couldn't build a robot to save his life. Fortunately, he has had nearly twenty years of experience writing books, magazine articles, and Web sites for kids. His work has been published by the American Museum of Natural History, TIME for Kids, Scholastic Inc., Sesame Workshop, and other organizations. He lives in New York City with his wife Ellen, two children, and a non-robotic dog.

GLOSSARY

This book is about a scientist who creates lifelike, futuristic machines called robots. Some of the words used in this book are "new" words created to describe how robots are designed and what they are programmed to do. To understand these words, it helps to break them down into smaller parts. For example, the word *animatronics* can be divided into two parts. The first part comes from *animate,* meaning to "put into action" or "make lively." The second part comes from *electronics,* the branch of physics that deals with the production, effects, and movement of electrons, especially in transistors, computers, and other devices. So an animatronic object is one that moves or acts by means of an electronic device and not on its own.

Here are some other word parts and their meanings that you may find useful as you read about robots: *auto,* meaning "self"; *micro,* meaning "small"; and *proto,* meaning "first."

For more information about these words, consult your dictionary.

actuators: motors that act like human muscles and provide a robot with a range of motion

algorithms: a set of rules or step-by-step procedures for solving problems, usually done by computer

animatronic: moved by wireless remote-control devices

artificial intelligence (AI): a branch of computer science that creates programs that empower machines to perform tasks and act in ways that usually require human or animal intelligence

autonomous robot: a machine that is programmed to function independently and make decisions on its own. For a robot to be considered truly autonomous, it must be able to react to its environment and make decisions based on its programming.

caliper: an instrument used to measure the diameter or thickness of an object

degrees of freedom: refers to the number of independent movements a robot can make. The more degrees of freedom a robot has, the more lifelike it may appear.

electromagnet: a core of magnetic material, like iron, which is wrapped by a coil of wire. Electric current is passed through the wire and magnetizes the core.

engineering: the science or profession that deals with designing, building, and managing engines, machines, sources of energy, roads and bridges, etc. To build robots, Cynthia uses her knowledge of two types of engineering: mechanical engineering and electrical engineering.

microchip: tiny machines that process information, make calculations, and control the flow of information; a chip containing thousands of integrated circuits.

microprocessor: the central processing unit for a computer; the computer's control center, memory bank, and calculator.

prototype: the first type or model of something

robot: a machine with moving parts and sensing devices that is controlled by a computer and which can carry out a series of tasks usually performed by humans. The word "robot" was created by Czech playwright Karel Capek and appeared in his 1921 play *R.U.R.* (Rossum's Universal Robots). The Czech word robota means "forced labor."

robotics: the study, design, production, and use of robots. A roboticist is a scientist who designs, programs, and experiments with robots.

sensors: devices that collect information about the area surrounding the robot, such as temperature, sounds, and sights

servo motor: a kind of actuator, in which an electric current is converted into mechanical movement

Metric Conversion Chart

When you know:	Multiply by:	To convert to:
Feet	0.30	Meters
Yards	0.91	Meters
Miles	1.61	Kilometers
Square feet	0.09	Square meters
Acres	0.40	Hectares
Pounds	0.45	Kilograms
Meters	3.28	Feet
Meters	1.09	Yards
Kilometers	0.62	Miles
Square meters	10.76	Square feet
Hectares	2.47	Acres
Kilograms	2.20	Pounds

FURTHER RESOURCES

Women's Adventures in Science on the Web

Now that you've met Cynthia Breazeal and learned about her work, are you wondering what it would be like to be a roboticist? How about a wildlife biologist, a planetary astronomer, or a forensic anthropologist? It's easy to find out. Just visit the *Women's Adventures in Science* Web site at **http://www.iWASwondering.org**. There you can live your own science adventure. Play games, enjoy comics, and practice being a scientist. While you're having fun, you'll also get to meet amazing women scientists who are changing our world.

BOOKS

Aylett, Ruth. *Robots: Bringing Intelligent Machines to Life.* Hauppage, New York: Barrons Educational Series, 2002. This overview of robotic science examines the dreams of AI pioneers from fifty years ago, exploring those that have come true and those that may happen in your lifetime. The book also offers fascinating discussions about some of the close connections between biology, engineering, and psychology.

McComb, Gordon. *Robot Builder's Bonanza.* New York: McGraw-Hill, 2001. If you're inspired to build a robot of your own, this resource will be extremely helpful. It's a how-to book jam-packed with lots of practical information on the electronics, mechanics, and programming of homemade robots. Included are illustrated plans for building 11 different robots.

Perry, Robert L. *Artificial Intelligence.* New York: Franklin Watts, 2000. This heavily illustrated book is a good introduction to the world of AI. It examines the different types of artificial intelligence, explains how it affects our daily lives, and imagines what AI of the future might be like.

Williams, Karl P. *Insectronics: Build Your Own Walking Robot.* New York: McGraw-Hill, 2003. Attention, budding roboticists. Here's another project book with step-by-step instructions for constructing and programming your own inexpensive six-legged 'bot. Your finished product will be a distant cousin of Attila and Hannibal, the two robots Cynthia built in the early 1990s.

WEB SITES

Engineer Girl: http://www.engineergirl.org
The National Academy of Engineering wants you to think about becoming an engineer. Learn about cool, well-paying engineering careers. Read fun stories about women engineers who are using their imaginations and ingenuity to solve problems and make the world a better place.

First LEGO® League: http://www.usfirst.org/jrobtcs/flego.htm
Young roboticists can participate in this contest sponsored by LEGO. You can have fun using sensors, motors, gears, and programmable LEGO "bricks," while you learn basic engineering and computer programming principles.

Learn About Robots: http://www.learnaboutrobots.com
This is an excellent site to get a quick update on robotics events that are in the news.

Low Life Labs: http://www.robotsandus.org/lobby
The Science Museum of Minnesota's Low Life Labs are the place to be if you want to explore a Robot Gallery (can you find Kismet?) or play interactive games in the Moving, Sensing, Thinking, and Being areas.

NASA Robotics: http://robotics.nasa.gov
http://robotics.jpl.nasa.gov/homepage.html
At the National Aeronautics and Space Administration, some of the brightest roboticists are developing spacecraft and rovers that are exploring Mars and beyond.

The Tech Museum of Innovation: http://www.thetech.org/robotics
Explore this online museum to learn about the history of robots. Use your computer to control your own virtual rover as it explores Earth or the Moon. Listen to a hot debate about ethics and robotics in the 21st century.

SELECTED BIBLIOGRAPHY

In addition to interviews with Cynthia Breazeal, her family, friends, and colleagues, the author did extensive reading and research to write this book. Here are some of the sources he consulted.

Breazeal, Cynthia L. *Designing Sociable Robots.* Cambridge, Massachusetts: MIT Press, 2002.

Brooks, Rodney A. *Flesh and Machines: How Robots Will Change Us.* New York: Pantheon Books, 2002.

Druin, Allison and James Hendler, eds. *Robots for Kids: Exploring New Technologies for Learning.* San Francisco: Morgan Kaufmann, 2000.

Menzel, Peter and Faith D'Aluisio. *Robo Sapiens: Evolution of a New Species.* Cambridge, Massachusetts: MIT Press, 2000.

Perkowitz, Sidney. *Digital People: From Bionic Humans to Androids.* Washington, DC: Joseph Henry Press, 2004.

INDEX

JHP Executive Editor: Stephen Mautner

Series Managing Editor: Terrell D. Smith

Designer: Francesca Moghari

Illustration research: Christine Hauser

Special contributors: Meredith DeSousa, Allan Fallow, Kate Nyquist Jerome, Mary Kalamaras, April Luehmann

Graphic design assistance: Michael Dudzik and Anne Rogers